Geraldine McCaughrean
The Odyssey
Illustrated by Victor G Ambrus

Oxford University Press
Oxford   New York   Toronto

For
Alexander Leslie Krasodomski Jones

Yearning for Home

The war lasted so very, very long. Then suddenly it was over in a flash of fire, a splash of blood and a trampling of horses. Men whose ships had rolled idly over a thousand tides in the bay of Troy mustered by the water's edge in groups.

There were many faces missing, many oars lacked a rower after ten years of war. But those who unfurled their sails, latched their oars over the oar-pins and set the tillers, were cheerful. Their masts were hung with tokens of victory and their holds were full of Trojan gold and wine. Best of all, they were going home.

Home! To wives they had not seen for ten years, to sons who had grown from boys into young men, to daughters who had grown from babies into beauties, to farms that had lain tangled and untended under ten hot summers. A few strokes of the oar and they would be home — all those men who had answered the call to war and mustered from every island and shore of the O-round ocean.

The long fast-ships were heaved off the sand and gravel and into deep water. Friends stood waist-deep in the sea waving and waving and waving.

'Till we meet again, Nestor!'

'Until we meet again, Menelaus!'

'Until we meet again, all you brave Myrmidons!'

'Safe journey, Odysseus!'

Odysseus felt the sand and gravel grate against the bottom of his ship. Then, with a rush of white water past the bow and the crack of his sail as it filled, he leaned on the tiller and turned his eyes away from the shoreline and the still-smoking ruins of Troy. He was going home to his three-island kingdom of Ithaca. His cockerel mascot crowed triumphantly on the stern rail.

Mustered behind his own fast, black ship, like cygnets behind their swan, were eleven others all manned by men of Ithaca, Cephalonia and wooded Zanthe. At first their rowing was ragged. Their oars beat out of time for lack of practice and their shoulders burned under the Trojan sun. But gradually they settled into a rhythm — a splash, a grunt and a sigh.

'Your son will be a big lad now, Captain,' said Polites.

'Eleven! Almost eleven! He was only a baby when I left Ithaca. A fine help I've been to his mother, leaving her all alone.'

'Ah, but such a lady, Captain! Such a lady as never knew the meaning of impatience!'

Odysseus looked into the distance with unfocused eyes. 'Indeed, yes, Polites. Such a woman.'

High in the window of Pelicata Palace, Penelope, Queen of Ithaca and wife of Odysseus, looked out across the wave-striped ocean. A dark shape caught her eye, far, far out across the sea. At once she was leaning out of the window and her hands were plunged into the unpruned vine which cloaked the palace wall. 'Odysseus! Odysseus!'

Her voice rang through the empty courtyards and tumbled over the cliff edge. Her son, Telemachus, stopped his game of archery and ran towards the house.

But it was only the shadow of a scudding cloud, and not a ship at all. Penelope pressed her cheek against the cold stone of the window frame and steadied her breathing. Behind her, Telemachus tumbled into the room. 'Is it him, Mama? Has Father come home from the war?'

Penelope turned away from the window, smiling. 'Not yet, Telemachus. I was mistaken. Not just yet.'

A breeze sprang up. The breezes braided themselves into a wind. The wind twisted itself into a gusting gale and the gale screwed itself into a frenzy. Odysseus' twelve ships were juggled by the waves: those on the crests and those in the troughs clashed sides as they rose and fell. The crews looked in terror at their comrades and saw them one moment against a sky crazed with lightning, the next in a valley of glazed black water, then enveloped in clouds of spray. They raised their oars but they were too slow to lower the sails which ripped in three. Their cloaks were so wrenched at by the wind that the cords half-throttled them. Two hundred voices called on the gods, and prayers skimmed like seagulls over the teeming sea. For nine days and nights they ate sopping bread and drank rainwater, cupping it out of the bilges with their hands.

'Land!'

'Where? I don't believe you!'

'There! There!'

'It's a cloud.'

'It's a reef!'

'It's an island!'

'We shall be driven past.'

'We shall be driven on!'

'We shall be broken up!'

'We shall be saved,' said Odysseus loudly and calmly, 'and the gods are to be thanked for it.'

The gods were indeed to be thanked. The storm died in an instant, and they found themselves on a sunlit beach of white sand. Strewn like flotsam, the twelve ships lay on their sides and the sea tickled their round bellies. The crews crawled up the sand, and most fell asleep on their hands and knees.

'Can we go and look for food?' asked Eurylochus.

'You don't want to rest?' said Odysseus in amazement.

'I've got a wife and six daughters to get home to, and I don't mean to keep them waiting any longer than need be, Captain. I've been away ten years already.'

'Very well. But go carefully. Take just twenty men with you: I don't want the islanders to think we are an invasion force . . . and don't get into any fights.'

Odysseus himself was anxious to inspect the boats for any damage. So Eurylochus took men and went inland in search of food and fresh water. The sinking sun wounded the sky. The night bruised it black. And still Eurylochus did not come back.

Odysseus waited until first light to begin the search. Leaving the ships well guarded, he took fifty men inland through the dense, luxurious trees. Velvety, succulent leaves stroked their faces. Sweet-smelling flowers drooped, heavy laden with nectar, and sprinkled their hair with pollen. There was a noise of water bubbling underground, and dark-eyed fawns peeped at them from between golden grasses.

'What danger could there be in a place like this?' whispered Polites at Odysseus' shoulder.

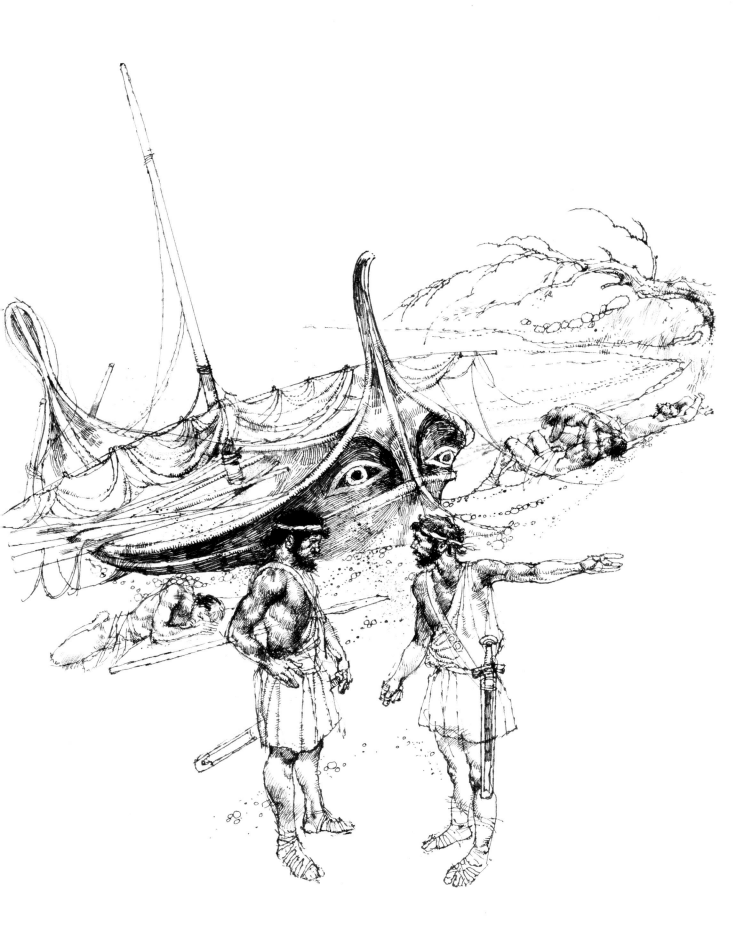

The King of Ithaca said nothing, but the hairs on the nape of his neck were lifting. No more than a mile along the green and shady path, they were dazzled by a clearing, bright with sunlit water. Round the lake stood a village. In the shade of the palm-leaf roofs, their sword-belts all unbuckled, lay Eurylochus and his twenty men as well as a pride of naked locals. The young native men and women all had long, thick hair which spilled over their shoulders and over the guests lying in the grass. They were plying their visitors with fruit from wooden bowls and, at the sight of Odysseus, leapt up smiling, and ran and took hold of the newcomers and dragged them towards the shade. Their hands were as brown as chestnuts and their skin as sticky as chestnut buds with the juice of the fruit. Their words were soft murmurs, hums like half-remembered tunes, and their mouths never once stopped smiling.

Eurylochus smiled, too. He smiled at Odysseus as at someone whose face was dimly familiar, and his words slurred a little when he said, 'Don't I know you? Come and have some of this fruit. There's plenty! Plenty! Taste it! You never tasted the like! Know you, don't I? Do I?'

He tossed a piece of fruit — a golden globe wrapped in a velvety skin — and Polites reached up to catch it. But Odysseus snatched the fruit out of the air and cast it into the pool. He whispered over his shoulder, 'Tell the men: no one is to touch the fruit.' He waved away the sticky brown hands that offered him the luscious food. Then he called out to Eurylochus, 'What of your wife and six daughters, my friend? Will you keep them waiting while you idle here?'

'Who? What? Sorry, friend, but I think you've got the wrong man . . . Wife? Daughters? Have some fruit. That's what you need — some fruit to set your brains straight.' And as Eurylochus spoke, the juice ran down his beard and stained his chest a sugary, crystalline gold.

Polites was alarmed. 'What's the matter with him, Captain? What's the matter with all of them?'

A native girl pressed a fruit against Odysseus' lips until he took a grip on her wrist and pushed it away. 'Have you never heard of the Lotus-Eaters, Polites?'

'The Lotus-Eaters?'

'Lotus-Eaters?'

'. . . –Eaters?'

The name echoed through the ranks of Odysseus' fifty men and their faces turned deathly white. Odysseus leapt up on to a poolside log. 'Courage, men! Your comrades have been eating the Lotus fruit. Their memories have melted and their wits have drowned in the treacherous juice. They care nothing now for us or for the families waiting for them. Are we to abandon them here? Or shall we save them from themselves? Close up your ears and seal up your lips, and help me carry them back to the ships!'

Round the pool they ran, pushing aside the fawning caresses of the villagers and overturning the bowls and baskets of Lotus fruit. They seized on their friends — two men to one — and dragged them to their feet.

'Leave us be! What are you doing? Get away! Who are you?' shrieked the Lotus-eating Greeks. 'You barbarians! Look, if it's the fruit you want, there's plenty for everyone! What are you doing? Where are you taking us? Leave us be! For pity's sake, don't take us away from the fruit!'

The further they were dragged away from the pool and down the shadowy path, the more desperately the advance-party struggled and pleaded and shrieked: 'The fruit! We must take the fruit! What are you doing? We can't leave without the fruit — we'd die! We'll all die without it! It's life! It's everything! Pity us! Don't make us leave the fruit!'

Shutting their ears and sealing their lips, Odysseus and his party of fifty men dragged their foolish friends down towards the sea, though their sandals kicked at the ground and their hands clutched at tree branches in terror. The Lotus-eating villagers pattered along behind making a murmured music with their whimpering. But as they got further from the grove where their beloved Lotus trees grew, they dropped away and ran back towards the village.

'Take some fruit! Please! A morsel of fruit, if you have a shred of pity in you,' begged Eurylochus.

'Should we, Captain?' asked Polites anxiously. 'We must have food if we're to row.'

But Odysseus forbade one Lotus fruit to be taken aboard, and the twelve ships were heaved into the surf as empty as they had come. 'What use would it be to row if we had forgotten where we were going?' he said. 'Tie the Lotus-eaters to their benches and don't untie them till this place is out of sight or they'll try to swim back.'

And so they would, but for the strong hemp that bound them and the determination of their friends who heaved on the shining oars.

At last their brains struggled free of the cloying nectar of the deadly fruit. They began to remember and to be ashamed. And, tight-bound to their benches, in the rolling bilges of the fast, black ships, they began to feel very seasick indeed after eating all that fruit.

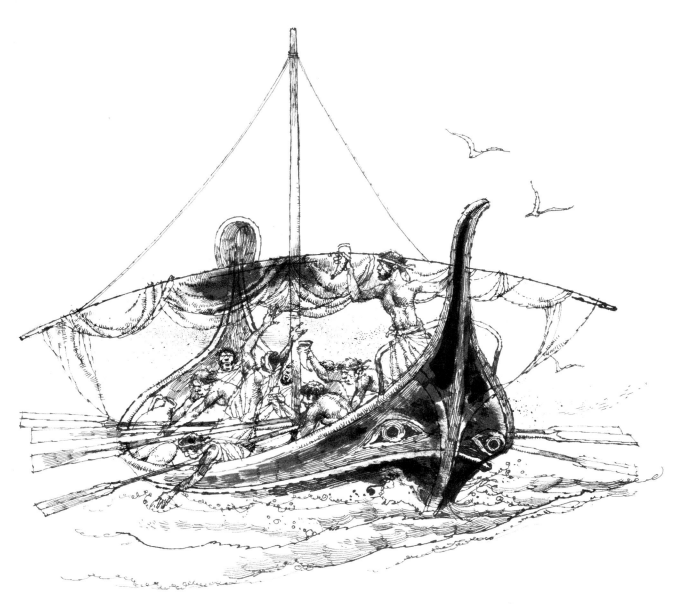

The Sea God's One-eyed Son

One thing there was in plenty: wine. Wine looted from Trojan cellars slopped in pointed earthenware amphoras rammed deep into heaps of sand in the stern of each ship. But as for food, there was not a bite left. Odysseus allowed his men to drink a sip of the Trojan wine in the hope that it would raise their spirits. But to his horror they immediately rolled into helpless drunkenness before slumping asleep on each other's shoulders. 'A little too strong,' he said to his cockerel mascot, and the cockerel shrugged its wings and fluffed out its feathers. The boats drifted, for want of rowing.

'Land!' shouted the look-out next day.

'Look, vines!'

'Olive trees!'

'Goats!'

'Let's go and load the boats right now!'

'Let's just tread warily and take what we are given,' said Odysseus. 'I shall take one ship's crew and make contact with the people living here. The rest of you moor by that little island off shore, and wait there till I send word that it's safe to join us.'

So one sole ship sailed up to the rocky mainland — past cliffs pitted with caves and planted with olive trees and corn. It is hard to judge size, looking inshore from a boat, or they might have wondered at the size of the cavernous cave dwellings or the height of the whiskered corn. Not a boat was moored in the bay, for the art of shipbuilding had not yet reached this remote outreach of the world. There was nobody about.

'Bring that amphora of wine,' said Odysseus. 'We may be able to trade it for food or give it as a gift if we are received graciously.' It took four strong men to carry the huge stone jar, slung between two oars by its looping lug handles.

They scrambled ashore and up a path to the nearest of the caves. The smell of cheese and sheep was overpowering as they entered. The wine-bearers set down the wine and propped the oars against the shadowy rear wall. As they did so, they fell over a huge, soft millstone of a cheese. 'Look at the size of this, Captain! Come on, let's take it and get out of here. It's food enough to last us as far as Ithaca.'

'What? Steal? When we could wait and be given it?' said Odysseus a little pompously. 'The laws of hospitality demand that our host give us food for our journey.'

The sinking sun shone in through the cave mouth, and as the night insects began to chirrup, the cliff-dwellers came back from pasturing their sheep and cattle inland. The crewmen could hear pebbles, dislodged from the pathways, tumble into the sea. Then the sheep arrived.

Sheep? They were the size of buffalo, fleecy as the bales of flax shipped in ones and twos on the great ships of Crete.

The sheep were delicate alongside their shepherd — a monstrous landmass of flesh and bone whose knuckles trailed in the dirt and whose mouth was a cave in itself. In the centre of his forehead, rimmed with rheumy lashes, gaped a single massive eye.

The Cyclops drove his sheep into the cave, rolled a boulder across the entrance to seal it, then revived the fire smouldering in the centre of the cave. As it flared up, it lit the oval staring faces of the astounded Greeks. The single eye gleamed as it fixed on each man in turn, and the Cyclops grinned. 'Hello, peoplings. Aren't you little?'

'Indeed, indeed. Poor miserable specimens come to admire the famous race of one-eyed giants,' said Odysseus (who was not just a hero and a king, but a diplomat).

The Cyclops had difficulty in hearing the small, piping voice. He cleaned one ear with his finger. 'Mmm. Two eyes. Almost repulsive. But I won't let it put me off. Me,

Polyphemus, I'll try anything once.' Reaching out, he picked up the fattest member of the crew and crammed him into that cavernous mouth.

13

It happened so fast. There was no scream, no shout of protest. When the second man was taken, the Greeks set up a clamour which shook the cliff, racing from side to side of the cave and beating with their fists on the rock.

'Sir!' cried Odysseus, struggling to keep the terror out of his voice. 'Where did you learn your manners? From the scum of Troy? Everyone knows that the gods frown on the man who shows unkindness to his guests!'

'Nobody frowns on Polyphemus,' said the Cyclops, tapping his hairy chest. 'My father's a god! I can do what I like.' He began counting them with an outstretched finger, and licking his hairy lips. 'Hmmm. Are there any more of you outside? Where did you come from? Did you come out of a hole in the ground like ants, or did you fly down from the sky?'

Made brave by fury, one man began to say, 'We came off the sea in warships, with swords and spears aboard, and yes, there are plenty of good men who . . .'

'. . . who would be here now if they had not been wrecked and drowned on the rocks,' said Odysseus quickly, to protect the five hundred waiting in ignorance on the little island off shore. (Better fifty should be lost than five hundred and fifty.)

'And who are you, skinny one?' said the Cyclops, walking his fingertips across the floor towards Odysseus.

For a moment he was tempted to throw back his head and declare, 'I am Odysseus, King of Ithaca, hero of Troy, whose deeds are spoken of by poets and whose kingdom encompasses Cephalonia and wooded Zanthe.' But Odysseus said instead, 'My family name is Atall, but I was the sorrow of my mother and my father's shame. So small am I of stature that my parents called me No-Body. What pride could you take in such a prisoner? Now roll back the boulder and let us go, or I won't give you the present I brought with me to your cave.'

'Present? What present? I like presents! Want a present! Give me a present! Give me, and I won't eat you, No-Body — promise, promise, promise!' said the Cyclops, thumping his fists on the floor.

'Very well. Since you ask so graciously. Men! Fetch out the amphora.'

Greeks crawled out from every corner of the cave, sobbing with fright. They had no wish to give the Cyclops so much as a smell of their sweat. But they trusted their captain and they heaved the stone winejar out of the shadows.

'Boo, is that all? I've got wine of my own,' snarled Polyphemus.

'Not like this, you haven't.'

So Polyphemus broke off the neck of the amphora and took a swig. 'Mmm. Fair.'

'The flavour settles to the bottom,' said Odysseus keenly. 'The first taste is good, but the last dregs are best.'

So Polyphemus drank it all, and he had to agree that the more he drank, the happier he grew until, when the amphora was empty, he was so happy that his brains

were like melted butter and his words as scrambled as eggs. ''Sgood present, No-Body. Likit. Polyphemush shleepnow. Alsho sheep shleepnow — baahaha! Morningtime eachew, No-Body!'

15

'Eachew?' enquired Odysseus, hoping he had misunderstood.

'No, shilly. Can't eat me! Eachew!'

'But you promised!' cried Odysseus indignantly.

'I lied,' said Polyphemus, with a beaming smile, before falling backwards unconscious.

For a moment there was silence — then a general rush for the doorway and a great heaving against the boulder.

It was futile. The men fell exhausted to their knees and wept openly.

'We're done for, Captain. It didn't work. Your plan didn't work.'

'My plan has only just begun,' said Odysseus from the rear of the cave where he had stood to watch them struggle with the boulder. 'Who has a knife with him? Help me work this oar to a point — and quickly! The drink won't keep our hospitable friend asleep for too many hours.'

With knives, and flints off the floor of the cave, they whittled at the rounded end of one of the oars by which they had carried the winejar. When it was worked to a point, they laid it in the dying embers of the fire. And when it was glowing hot and ready to burst into flames, they tempered it, hard as metal, by splashing it with milk from the ewe-sheep. Again they laid it in the fire. By the time it was glowing white-hot, the darkest, latest hour of the night had begun, and the monstrous Cyclops was beginning to stir.

Men who had charged the brass gates of Troy once more stood side by side, the sharpened oar resting on their shoulders like a battering ram. Odysseus was nearest the glowing point. He gave the word to run forward. He aimed the oar. He guided the point into the opening eye of the waking Cyclops. But he and all his men fell back from the noise which followed.

Polyphemus arched his back and clawed at them and at the pain in his head. He took hold of the quivering oar, and wrenched it out of his tormented face and hurled it. The sheep scattered in terror. The Greeks threw themselves on their faces and prayed to the gods. The Cyclops' screams clamoured in the cave like the clapper in a bell — rattled the cave in its cliff-face. Landslides crumbled into the sea below.

Other Cyclopses were brought from their beds — men and women all as vast as trees — crashing about in the outer darkness, hurrying to the aid of their neighbour. 'What is it, Polyphemus? Who's in there with you?'

The cry came back: 'No-Body! No-Body Atall has hurt me. No-Body Atall is in here. Oh, someone tell my father! No-Body Atall has blinded me!'

The Cyclopses peered at each other in the moonlessness. 'Well, that's all right then. A nightmare obviously. We're glad to hear it, Polyphemus! Peace be on your eyelid till morning!' And away they went, a little bad-tempered at having been roused up for no purpose.

When Polyphemus heard them go, he lapsed into a terrible silence, staring about him at the unutterable darkness of his everlasting night. At last he said, 'Your plan has failed, No-Body. I am not dead. But you and your comrades will never leave this cave alive.'

Outside, the night sky turned pale with fear. It was morning. But no sunlight creeping in round the massive boulder told Polyphemus that it was daytime — only the bleating of his sheep. 'Oh, my woolly ones! You want to be out in the daylight. Of course you do. Don't I know, better than any of you, how to long for the sunlight? It's gone! I'll never see it again. I'll never see anything again. Blind! Oh, you gods! To be blind for ever! The sea's blue is nothing but a noise. The grass's green is nothing but a wetness underfoot. Oh my dear, happy, ignorant little sheep! If only you could speak and tell me where those Greeks are hiding. I'd pull them like wishbones. I'd have them die twenty times over in the killing of them!'

Feeling the contours of the familiar boulder which stoppered up his cave, he put his shoulder to it and rolled it aside. But he sat himself in the very centre of the doorway, with his hands spread to either side, so that no loathsome Greek should pass him alive. The sun warmed his back and his sheep pushed forward, bleating. 'Slowly, now. Quietly, my little loves,' said the Cyclops tenderly. 'It would never do for the villains to get by me by clinging to your fleece.' And he felt along their backs and along their fleecy sides before he let them pass.

Little did he realize that Odysseus had roped the ewes together in threes, and that underneath each centre sheep a man clung on for his life.

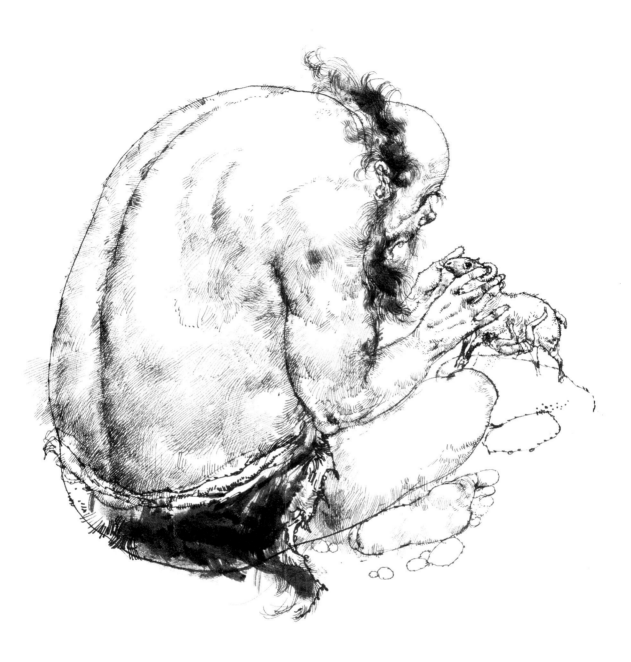

Soon all the sheep but one had passed Polyphemus. Only the big old ram remained, with Odysseus himself clinging beneath its belly. As it passed the Cyclops, he took hold of its head in his two great hands, and wept from his lost eye. 'Oh, my old mate. My dear old friend and companion. Would to the gods you could speak and describe the beauty and the wildness of the world. What use am I to you now? Can I milk your ewes or guide you to the pasturelands? I shall have to give you to my worthless neighbours — the scum who left me in agony last night and never came to my aid. Oh, woolly one! I'm sorry! You'll never, never know how sorry I am!'

19

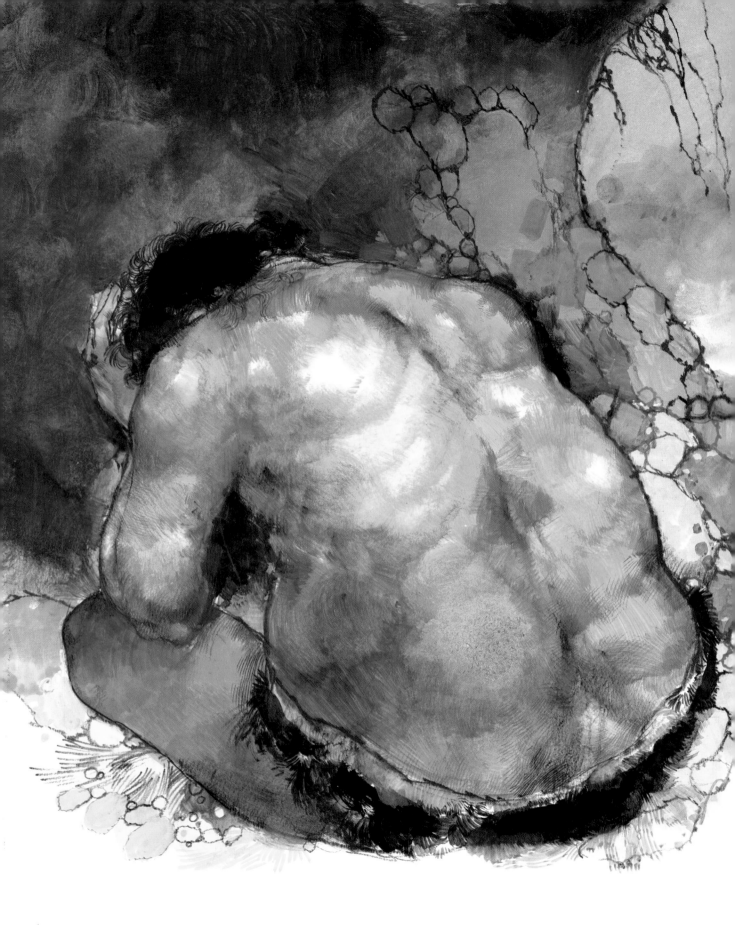

At last he let the ram pass by, and Odysseus dropped on to his back on the rough cliff path. He took to his heels and ran after his men, gathered by now on the beach below. They were wrestling the sheep into their fast, black ship: food for the voyage.

They pushed off. They bent across their oars. The sea rose white beneath the prow. Their course lay past the cliff pitted with caves — right below the cave-mouth where Polyphemus sat feeling about for his enemies. The memory of his two dead companions galled Odysseus: he could not unfix his eyes from the huge, hairy back of the weeping Cyclops. All at once he got to his feet and roared, '*I am Odysseus, Polyphemus. I am Odysseus, hero of Troy, and my kingdom encompasses Ithaca, Cephalonia and wooded Zanthe! It was I who blinded you, and the poets will one day praise me for it in songs of sixty verses!*'

21

The men at their oars stared at him in disbelief. Even his mascot, his own cockerel, pecked him in the arm. But Odysseus was unrepentant. 'What harm can it do?' he blared. 'There's no one but a blind Cyclops to hear me — ha, ha, ha!'

Polyphemus heard the taunt and rose to his knees, then to his feet. He cocked his ear towards the sound of Odysseus' voice. He picked up the boulder from the opening of his cave and raised it over his head. Before he let it go, he raised his blinded face to the heat of the sun and bawled, 'Father! You god of the oceans! Poseidon, god of the sea — hear my curse! See what Odysseus, King of Ithaca, has done to your son! Hate him with all the heat of the Earth's core — as I do! Hate him with all the unforgivingness of the Earth's icy peaks—as I do! Curse him as I curse him! Avenge me, for I am powerless to be avenged!' And he hurled the boulder.

It hit the water a fraction behind the stern post, and the wave it raised lifted the boat like a hand and thrust it forward, gouging a furrow through the sea. Headlong they hit the little offshore island. Keel-long the boat split, spilling the rowers on to the silver shingle. Odysseus, as he rolled clear, laughed out loud and kicked his feet in the air. 'So much for the curses of a Cyclops!' he snorted as his crew of five hundred and more gathered round him, slaughtering the captured sheep.

But of those who had escaped from the cave, not one laughed and not one congratulated him. Two of their comrades were dead — eaten by the Cyclops — and Polyphemus had cursed them.

Odysseus scowled and lay on his back, looking up at the sunny morning sky. 'Poseidon, did he say?' whispered a voice over his heart. 'Are we to be cursed by Poseidon, the god of the sea?'

And somewhere in the ocean's well, the cries of Polyphemus set the electric rays trembling and the jaws of clams agape. 'Polyphemus is blinded!' they cried. 'A curse on Odysseus and on all his men!'

The Brass Island and the Bag of Winds

At first it was just a dazzle on the horizon, a flash too bright for the eyes. Then they began to make out its shape.

'Land!' cried the look-out.

'No, a ship, surely?'

'An island.'

'A city!'

Aeolia was all of those things. It rose out of the sea like a great inverted brass bucket — floating, bobbing, bound about with brass cliffs as high as the walls of Troy. Only a single white line of caked sea-salt stained its shining, polished sides. There was not a ladder, a rung or a foothold. As they sailed alongside, they could see their own faces — a strange sight after ten years spent living in battle tents. While they groomed their beards and hair, Odysseus cupped his hands about his mouth and hailed the people of Aeolia.

Right beside him, a wirework basket, woven in the shape of a housemartin's nest, dropped down on the end of a brass chain. An arm was beckoning from the rim of the high brass wall. Without a moment's hesitation, Odysseus leapt inside. 'At the first sign of trouble, push off and get well away. Polites is in command if I don't return.' As he spoke, the basket was hauled up.

When he reached the top, two hands helped him out of the basket. They were soft hands, heavy with jewels. 'Welcome! Welcome to Aeolia, stranger. Come and dine with me and my family. Shall I fetch up your men or send food and drink down to them? The rules of hospitality command that I give you everything you need.'

'Your kindness does you credit, sir,' said Odysseus, and introduced himself modestly enough.

'Odysseus? But I have heard so much about you! Every ship that passes brings some new news of Troy and its heroes, and your name is always mentioned. But the war is over. What are you doing so far from your three-island kingdom?' The King of Aeolia was hungry for news: he gobbled it down like food and drink, and Odysseus quickly understood why.

In the dining hall, all the people of Aeolia were gathered: the King's Queen, his six sons, six daughters, and a handful of servants. Like two teams of chess-men, they faced each other across the shiny tiled floor, and a tinkling music resounded over their heads where strings of seashells jostled in the breeze.

Odysseus sent word to his men that they were in safe hands. But seeing the great shortage of chairs and the one dining table, he insisted they stay in the ships below to eat and drink all the good things King Aeolus sent down to them. How they dined! All evening and all night they ate, until the eleven black ships sat low in the water and the sailors slept over their oars.

High above, there was no sleep for Odysseus. To satisfy the endless curiosity of the King, he had to recount all his exploits of bravery in the wars, all his adventures since leaving Ithaca. The ruler of this floating castle, this drifting kingdom, had never set foot on the shore of the O-round sea, and he lived for the gossip of sailors.

Odysseus said, 'You must come to Ithaca some day and let me repay your hospitality.' But then the King's features froze, and all of a sudden the beautiful brass city of Aeolia seemed like a prison.

'Oh, we *never* leave the city. We have everything we need here. I've married my sons to my daughters so that they need never leave home, and travellers like you tell us stories of the world. What more do we need? . . . Enough! Come with me. I have a present for you.'

He took Odysseus by the wrist and led him through brass corridors to a steely room locked with a golden key. Inside it was a single bag fastened with seven cords. But it was a bag such as the god Helios might have sewn out of the hides of his own cattle — a skin bag with seven seams, neither round nor square, but writhing slightly where it lay. Something was inside it. While servants carried the bag up to the roof of Aeolia, the King explained. 'Last week, Zeus the Almighty, Father of all the gods, quarrelled with Poseidon, the Sea God. To punish him, Zeus confiscated from him the eight winds of the world, and put them in my safe-keeping for five days. The five days are up, but before I give Poseidon back his winds, why don't I lend them to you, my dear Odysseus? I shall set just one free — the soft westerly breeze that will carry you home to Ithaca. If you keep all the rest safely penned up in the bag, they can't hold you back or endanger you with storms and rough seas.'

Up on the roof, he eased the cords a fraction and thrust in his arm, right up to the elbow. His clothes billowed: he was almost lifted off the floor. But at last he pulled out a white rag of a thing — a corner of westerly breeze. Seven servants hauled on the seven cords to shut the bag again.

'Go with my blessing, Odysseus, and with a full sail. No need to row — only to keep a straight tiller and watch out for the shores of your homeland.'

Odysseus was lowered down with the bag into his own ship, where he found his men struggling to raise the sails and catch the favourable breezes. 'No hurry, men.

There's plenty to carry us all the way home. Stand away from the tiller. I mean to steer this little fleet of ours all the way into the harbour below Pelicata Palace!'

'What's in the bag, Captain?'

'Treasure!' declared Odysseus delightedly. 'The best present any host could have given to a weary traveller. Nobody touch it, you hear?'

And foolishly, that was all he said.

He stood at the tiller with one foot resting on the wriggling bag of winds, and he looked into the distance, thinking of his wife and his little son and his three-island home.

The royal family of Aeolia waved from the parapets of their brass home — a lonely sight for all their wealth — drifting forever in the heart of the sea. Soon Aeolia was nothing but a flash too bright for the eye, away on the distant horizon.

'What is it, do you suppose?' whispered Eurylochus to the man alongside him. (The man shrugged.) 'He said it was treasure — and we're not to lay hands on it. That's how kings share their winnings, is it? No wonder he left us down in the ships while he went up into that brass treasure-chest of a place. The King gave him some gold, or jewels or some such, and he's keeping it to himself. Ten years we fought alongside him, and this is how he repays us. Look at him: he won't take his foot off that sack.' Again the man beside Eurylochus shrugged, then, resting his forehead on his arms, he went to sleep across his oar.

But for ten days and nights Eurylochus kept wondering, kept asking his unanswerable questions of the men around him. He did not dare to ask Odysseus outright. There was a better way to find out the truth. He had only to wait, and Odysseus would fall asleep; night and day he stood at the tiller with his foot on the sack, and would not leave it even to snatch an hour's rest. The rest of the crews were well-rested, however, when the hills of wooded Zanthe came into view, and seaweed off the beaches of Cephalonia and Ithaca itself drifted by. People on the foreshore mending nets shielded their eyes and looked out to sea at the approaching ships. Then Odysseus felt safe at last. 'Someone take this tiller. I must sleep. I can't stay awake another moment.'

Eurylochus leapt the length of the boat, all smiles, all helpfulness. 'Let me, Captain.' He took the tiller and watched Odysseus curl up against the cowhide flank of the bag. And while other men were standing up, exclaiming and pointing out the familiar landmarks of home, Eurylochus eased just one of the seven fastening cords beside Odysseus' sleeping head.

The people on the shore rubbed their eyes. They thought they had spotted a fleet of black ships, but now there was nothing but a funnel of water gouged out of the sea like the core of an apple. Above it the sky filled with rainclouds, and the whole plain of the ocean was crumpled into mountainous waves. Spray blanked out the horizon.

The eleven black ships, like splintered shards, spun away through the mist, driven hither and thither.

Odysseus was thrown off the bag as the winds inside it writhed free. Eurylochus was bowled heels-over-head to the very prow of the ship and left clinging to the figurehead. The cockerel was buffeted high into the air, and the oars were wrenched about like the legs of a dying insect, as the fleet was driven at the mercy of eight winds.

The hot south wind burned their skin, and the cold north wind froze their hands to the ropes as they fought to save the masts. The ripped sails enveloped men like shrouds and carried them over the side.

As, in the seige of Troy, the warrior Achilles killed Prince Hector and dragged him by his heels, behind a chariot, three times round the walls of Troy, so the winds dragged Odysseus' fleet three times round the ocean. Somewhere in the darkness below, Poseidon the Sea God felt his powers restored to him. He plucked up his eight winds, replaced them in his quiver and said, 'Now listen, Polyphemus, my ugly son, and hear the first note of my revenge on Odysseus.' And his hand, knotted with veins of purple water, took hold of the spinning fleet and hurled it — where? . . . Against the brass wall of Aeolia.

Like hammers against a gong, the ships struck 'eleven' against the towering brazen walls, and the men were jolted against their own horrified reflections. Their sweating hands made greasy prints on the yellow metal. Their panting breath slubbered the shine, and their fists beat — clang clang clang — on the impenetrable walls, and brought the royal family to peer down from the roof.

Odysseus called up, 'Lower a basket and let me tell you the foolishness that brought us back here! Give us shelter in your friendly home!'

A golden coin struck him in his upturned face, and others hit the men around him and made them cry out.

'Get away, Odysseus of Ithaca,' came the reply. 'Get away from my spotless kingdom before the gods mistake me for a friend of yours. It's plain to me that you have offended the immortals. You are a smell in the nose of Heaven that must be sneezed away. I am a god-fearing man: my wife and children are god-fearing people. I won't help an enemy of the gods. I won't. Now get away. Let go!' And coins and sharp-cut jewels rained down in their faces and drove them down beneath the shelter of their benches.

The eleven ships drifted unsteered, rolling in the swell. On the bottom boards of the boats, five hundred and more men prayed to Athene, goddess of war, to Helios the god of the sun, to Hera the mother of the gods, and to Zeus the Almighty himself. (They did not pray to Poseidon, for in their heart-of-hearts they knew that his back was turned on them.) When they first saw the curving bays of Laestrygonia reach out to them like welcoming arms, Odysseus believed that their prayers had been answered.

Two curving headlands circled a natural lagoon of green water so clear that fishing pots were visible several fathoms below on the sand. The entrance was so narrow that the ships went in single file, and Odysseus tied his own by a single rope, just outside. The ships already berthed in the lagoon were marvellous. They made the Greek fleet look like children's canoes. Odysseus was so intrigued that he walked the length of the spit with his eyes fastened on them, and bumped into a tree.

It was a brown tree with fine blonde hair coating the trunk, and roots which splayed out in only one direction and ended in . . . 'Toes?' said Odysseus, and looked up. A jolly, smiling girl leaned down and picked him up in the palm of her hand.

She examined him on all sides, lifting his tunic with one finger to delight in his miniature underwear. Taking the end of her blonde plait, she brushed him down with it, still beaming with joy. 'Look, Mother! Look what I've found! There are lots!'

Gathering up as many Greeks as she could carry, she ran along the harbour promenade with them cradled in one arm like so many peg dolls. She beckoned the others to follow. 'Come along, little ones!' she called, and clucked and whistled and made little kissing noises as if to encourage them. They followed, spellbound with horror, and because their captain was tucked snug in the crook of her elbow.

Her mother was equally pleased to see the visitors. She dwarfed her daughter, and her voice was as loud as a landslide. 'Wait till your father sees who you've brought to dinner, my darling dear.'

Her husband, King Lamus of Laestrygonia, dwarfed his little wife, and ships far out at sea thought that his white hair was snow on the mountains. They mistook his palace for a mountain range, for its buttressed walls reached so high that eagles nested in the eaves and clouds billowed like curtains in the arching attic windows. He was delighted by the guests his daughter had brought home to dinner.

For King Lamus and all his Laestrygonians were cannibals.

With great good humour, he crammed two sailors into his mouth and crunched on their bones and picked their leather clothing from between his teeth. Odysseus, driving his short sword into the elbow of the giant princess, heard her squeal and found himself plunging to the ground. Once there, he picked himself up and ran — leaping, zigzagging, somersaulting down the palace steps and on to the harbour promenade. As vulnerable and small as ants, he and his men swarmed back towards their ships, dodging and ducking the scooping fists of the giants.

Somewhere in the belfry of the palace, an alarm bell began to clang, and out of houses high as hills came all the citizens of Laestrygonia. A cheerful, smiling people they were, with ruddy faces, thanks to their meaty diet. A harvest of five hundred men was a rare treat, though, even for them — a harvest-festival of flesh. They stamped and ate and scooped and ate and fished with their hands into the clear water to scoop out those that escaped them on dry land. It was like ducking for apples. The Laestrygonians laughed out loud at the sport (even though their mouths were full). Some of the laughable little men even managed to reach their ridiculous little boats and axe the mooring ropes as if they might escape to sea! How absurd.

The Laestrygonians merely took the boat-prows in finger and thumb and twisted them over, tipping men and oars and amphoras and sheep, storm-broken rigging and Trojan treasure into the clear, green water. Then it was an easy matter with tridents and throwing harpoons, to fish for the tender little oddities. They ate them raw, with only salt seawater for seasoning.

King Lamus was first to notice the column of fleet-footed creatures running along the headland to the mouth of the harbour. He waded into the water, pointing. His citizens and subjects went after Odysseus and the fifty others who were heading for the harbour-mouth. One or two of the stragglers were picked off with harpoons, and each success was greeted with cheers from all sides of the lagoon. But a vexing number reached the tip of the mole and leapt off it into the outer sea. Too late, King Lamus saw that a single ship was moored outside the lagoon, that the creatures had jumped into it and were already bent over their foolish little oars.

Odysseus slashed the bow line with his sword, and the fast-ship leapt through the swell with such a jerk that he lost his footing and sprawled on the deck. As he did so, a Laestrygonian trident flew over his head and impaled the first oarsman.

'The gods forgive me,' said Odysseus softly. 'One ship left out of twelve! We are cursed indeed!'

The Pig-Woman

'I don't know if we shall ever reach home again,' said Odysseus to his remaining crew. 'All I know is this: a man's fate is decided on the day he's born, and we shan't any of us go down dead to the Underworld a day before our appointed time. So stop that crying. Two days is time enough to spend crying. We have done what is expected. We have called the names of our dead friends three times across the ocean so that they shan't go down nameless into the Underworld. Now we must turn our minds to the wives and children waiting for us at home. We must find out where in the world's round blue sea we are.'

The men stirred, wiping their tear-streaked faces. They looked up at the sky, but it was one low, white fret of mist and even the sun did not show through. At night there were no stars. Navigation was out of the question.

35

'Land!' called the look-out.

'Where? How can you tell?'

'I can smell it. I can hear waves breaking on a beach.'

And so they beached, not knowing whether this was mainland or island. When the sky cleared that night, the constellations were unfamiliar — strange beasts prowling an unfamiliar sky. The men shivered at the thought of meeting yet another race of cannibals, Lotus-Eaters, or monsters.

A path led inland from the rear of the beach and, keeping close together, they followed it. Odysseus glimpsed a stag between the trees and went in pursuit, killing it with a single arrow and carrying it back to the ship across his shoulders. Laying it down, he ran after his men to catch them up. But he could not do so before they reached the farmhouse.

It was a shady place of irrigation streams, springs and bougainvillaeas. A walkway of overarching vines led to a curved door of brass. As the men approached, the doors swung open in welcome, ushering them into the shade of a dining hall. A table was laid there, and a woman with lilac eyes stood calling them by pleasant names. Her plaits of brown and golden hair were like the hawsers of a beautiful white ship all hung with flags.

Odysseus saw the last of his men go inside. By running, he could have stopped the brass doors closing: he could have gone in with his men and sat down to dinner with them. But for some reason he hung back. His feet would not hurry him across the lawn of white moly flowers that silenced his tread. Instead, he sheared off around the house and walked about in the farmyard, looking into the pig-pens, trying to quieten the unaccountable beating of his heart.

Then he crept back to the window and watched what was happening inside.

The woman with lilac eyes had seated his men at a table spread with white linen. She gave them fresh warm bread and bowls of tzatziki, with whey to drink, and peeled fruit, and parsley in soft cheese. At least, it looked rather like parsley, that sprinkling of green. She gave them wine, too, and more wine, and then . . .

Odysseus dared not close his eyes, although what he saw was too horrible for one pair of eyes to witness alone.

As they ate, she walked round the table — round the backs of their chairs. She carried a willow wand, plaited like her hair, and as she passed each man, she knocked him about the head with it — gently, as if she were teasing.

But it was not teasing. Each man's legs at once began to shrink, until he rolled on his haunches and could not keep his balance in the chair. He would reach out to steady himself, but his arms had shrunk, too. No hands at the end. Only hoofs. And the men fell out of their chairs and into their suppers — their snouts in their suppers.

Snouts, ears, trotters—and curly tails splitting their tunics. They were pigs, every man of them. Pigs! The animals in the pig-pens in the yard set up a nightmare squealing, and racketed against the bars.

36

Odysseus ran back round the house, thinking to break down the brass doors, thinking to cut the long-haired woman in pieces for what she had done. But once again his feet would not speed him across the soft lawn of white moly flowers. It was some time later that he knocked on the door and was welcomed inside by the sorceress Circe.

Her brown and golden plaits trembled a little at the sight of Odysseus (for although he was a small man of stocky frame, his hair curled like clematis, and his eyes were very brown). Nevertheless, she beckoned him indoors to a single place laid at the long, linen-covered table. Her accent, when she spoke, sounded as though she had been born in the very shadow of Pelicata Palace. But that was magic, merest magic, Odysseus told himself.

'You are late,' she said. 'Your friends have already dined and gone to walk in the gardens.'

The chair she sat him in was carved with flowers and birds. The wine she handed him smelled sweeter than the evening primrose. The meal she laid before him — tzatziki and olives, peeled fruit and cheeses, wine, honey and fresh warm bread, reminded him of meals taken with his Queen, Penelope, in the shade of Pelicata's vineyard. But that was magic, merest magic, he told himself.

He drank down the wine. He ate the food — even the small green herb like parsley — and then he sat back and wiped his beard with a linen napkin.

She struck him hard with the willow wand. It raised a red line across his cheek. She said, 'Handsome or not, no foul man must be allowed to keep his shape on Circe's magic isle. Now get to the pigsty with the rest.'

'No,' said Odysseus, putting his feet up on the table.

'I said . . .'

'And I said "no".' He took out his sword and calmly thumbed the blade. 'You see, lady, on the hour I was born some friendly god or goddess cloaked my heartbeat with wisdom. That same wisdom taught me that the little white moly flower is the antidote to many a magic potion and poison.' And he spat out the petals he had pouched in his cheeks. 'Now, before I kill you, you have one last magic spell to speak, madam. Give me back my men, or you will indeed be sorry that you were ever born.'

'Odysseus!'

He was taken aback by the sound of his own name in the mouth of a complete stranger. Circe sank to her knees in front of him and laid her head on his knees. 'Odysseus! On the day I was born, a prophecy was written: that one day I should be overpowered by Odysseus, King of Ithaca. I cannot choose but love you: it is my fate! I pray you can find it in your heart to love me just a little in return!' And she began kissing his knees passionately.

'Lady! Please! You have just turned all my comrades into pigs! Love is not the word to describe what I feel!'

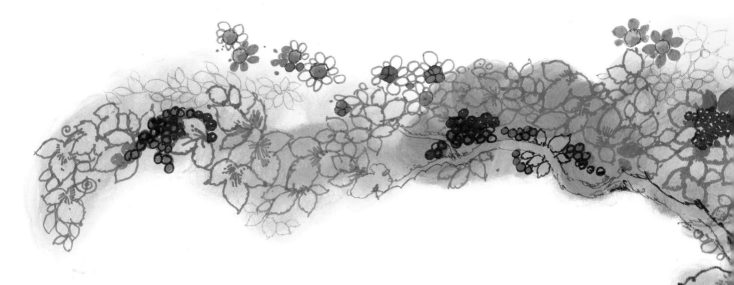

At that she took him by the wrist and pulled him to his feet, rushing him into the yard with her willow wand outstretched ahead of her. She fumbled with the gate of the sty and, as each pig dashed out in a frenzy of squealing, she rapped it across its bristly back.

A moment later forty-five shivering men crouched whimpering in the yard, their hands and feet all clogged with mud, and pigswill clinging to their beards. They would have mobbed Circe and killed her where she stood but for a terror of her willow wand. Circe, meanwhile, was kissing Odysseus' curling hair.

He dodged away, blushing. 'Next, lady, you can tell me the latitude of your island and where I should look for the Dog Star. This quarter of the sea is strange to me and I must set course for Ithaca.'

Circe knotted her hands in her long shining plaits and burst into tears. 'Oh, don't leave me, Odysseus! Stay with me! A hundred years I've waited for you, and though it is my fate that I should love you and lose you, I won't let you go so soon. I won't! I won't!'

Odysseus was in a quandary. His solitary ship was in tatters. His men were exhausted. Out there, beyond the bowers of Circe's magic garden, the god Poseidon sat seething on the sea's bed. Even so, Ithaca was waiting — a kingless kingdom and a lonely queen. 'I insist, madam. Tell me how I must steer to reach Ithaca.'

'I can't,' said Circe. 'The gods have forbidden me to help you.'

Odysseus gave a cry of exasperation and turned away, making for his ship.

'Wait! If you stay with me one month, I'll tell you how to find out that and more! I'll send you to someone who knows the past, the future, and the truth, and will tell you all three!'

'A month?'

'A little month,' urged Circe, and her white hands were already unbuckling his swordbelt and loosening his tunic.

Alive Among the Dead

One month became a season. A season lengthened to a year. And only then did Odysseus think again of his three-island kingdom. Life with Circe was as sweet as Lotus fruit: it tended to make a man forget his home and family. Then his best friend, Polites, came to him and said, 'Poseidon's memory may be long or it may be short, but yours has failed you altogether if you forget your beautiful Queen. Your men have wives and children, too, and we have been gone from them now for more than eleven years!'

So Odysseus went to Circe and held her in his arms apologetically, and said, 'It's time to go. You promised me a year ago to send me where I can learn the route home and the secret of things to come. Who is this oracle? Where will I find him?'

Circe bit her lower lip and clenched her fists. 'Very well. I will tell you. It was written when I was born that I would love you and lose you. But you won't like the directions I give you. You may not dare to follow them.'

'Not dare, lady! I am Odysseus of Ithaca, hero of Troy, whose exploits . . .'

'Yes, yes. All right. So be it. Your path lies through the shadows of the Underworld. There, among the spirits of the dead, you will find the oracle Teiresias. He can tell you what is past and what is to come, and what is true, besides.'

'No!' cried Odysseus. He put his hands over his ears and he screwed shut his eyes. 'No! No! No! Unsay it, Circe! See how I tremble! See how the sweat breaks out on my face! Unsay it, Circe, or you'll make a coward of a man who has looked death in the eyes fifty times and never flinched! Go down to the Underworld before my time? Rub cheeks with ghosts in the bottomless dark? Zeus! A man's heart would shake itself to pieces! No! Never! No!'

Circe was silent and her eyes delighted in the thought that Odysseus would stay with her now for ever. He threw himself down on her white couch and howled like a wolf for an hour or more. Then he stood up, took three deep breaths, squared his shoulders, and set off for the beach where his men were sleeping by the ships.

'Aboard, men! Aboard now and I shall take the tiller! Circe the Sorceress has told me our route home, and it's time to set sail.'

As their red-prowed fast, black ship scraped its hull through the white sand, and its dry planks swelled at the touch of the sea, Circe ran down the beach and plunged knee-deep into the surf. 'The gods keep you safe, Odysseus! Place your keel in the path of the sinking sun. Then River Ocean will draw you on without the need for oars. Bind the tiller and make sacrifice to Hades, God of the Dead.'

Her voice roused Elpenor.

A quick-footed but slow-witted fellow, Elpenor had gone to sleep on the flat roof of Circe's house, and he had missed the ship. The sun shone hot on him as he slept. A red haze stuffed up his eyes when he opened them. He reached for the ladder by which he had climbed up to the roof, but stepped off into thin air. With a startled cry, he fell head-first and, in hitting the ground, snapped his neck.

'Where's Elpenor?' asked Palmides. 'He's not here, Captain. Should we turn back for him?'

But there was to be no turning back. The keel of the red-prowed ship had already been seized by the River Ocean — a current which lay beneath the orange path of the setting sun. Though the rowers shipped their oars, their ship picked up more and more speed. Water raced under the bow with a whispering hiss, and the men's delight at heading home changed to a nervous uneasiness. 'Where *are* we going, Captain?' said Eurylochus. 'Where has that witch directed us to go?'

'To Hell,' said Odysseus. 'To a place no living man has seen before. To the Underworld. To the Kingdom of Hades, God of the Dead. To the spirit world. *To Hell.*'

The shell of the night sky shrank suddenly to the size of a black cave, and all the stars went out. The current sucked the ship deep into the cave, and the men's sobbing echoed back off unseen walls. When they stretched out their hands, soft, slimy plants or creatures recoiled from their touch. Mouths sucked at their fingers. Every man crawled under his bench and cowered there, moaning and complaining that his life had been cut short.

Then the keel jolted aground, and white hands curled over the prow and pulled the ship on to some shallow, unseen beach. Faces floated like jellyfish through the dark, cold air, and brushed against them as they set foot at last in the Kingdom of the Dead.

'Elpenor!'

It was the first face they saw with any plainness — a shred of a face, with sad eyes and an 'O' for a mouth. 'How did you get here ahead of us?' But as his friend Palmides rushed forward to embrace Elpenor, he clasped only a wedge of clammy air. 'Elpenor! What's happened to you?'

'My body lies unburied on the island of Circe,' wailed Elpenor (though his voice was almost too small to hear). 'If you had turned back . . . if only you had cared enough to turn back and look for me and give me decent burial! But I came here nameless and the spirits won't speak to me, because I had no proper funeral. Oh, comrades! Stay here with me. Don't leave, will you, or I shall be alone and unknown for ever!'

'We shall go back and give your body decent burial,' called Odysseus as the face was blown, by subterranean draughts, away along a corridor of darkness.

They walked on, their sandals making no sound on the spongy slime underfoot. Each few moments one of them would give a startled cry as he recognized a relation or friend long since dead. Many heroes who had died in the Trojan Wars hailed them from out of the shadows.

Worse was in store for Odysseus. He glimpsed his own mother, like a streak of moonlight, in a black garden of colourless flowers. So she would not be waiting to welcome him beneath the shady vines of Pelicata Palace. He had stayed away too long to meet her again in the land of the living. She greeted him mournfully. 'My son. Did you die in the wars or were you drowned on the voyage home? Have you only just arrived? I hope your friends gave you decent burial.'

'But, Mother, I'm not dead,' protested Odysseus. 'I'm here because my travels have brought me here. My time hasn't come yet to die and live here with you.'

'What travels, son? You mean to say you haven't been home yet to Pelicata Palace? The dead warriors say that the war finished long ago. Why so slow? How will your poor Penelope fend off the suitors?'

'Suitors? *What suitors?*' demanded Odysseus.

'A rich and beautiful widow will attract many men to woo her, my dear son. Even when I died, a year ago, the shores of Ithaca were bright with coloured boats. Soon Penelope will be forced to choose a new husband and a new King of Ithaca. She must surely have given you up for dead.'

'But she's not a widow! I'm not dead! I'm alive! This is terrible! Where's Teiresias? Where's the Oracle? I must get home instantly!'

'You came here to see me, and yet you delay even now making idle conversation with your friends and relations.' The heavy darkness was prised back by a single beam of light — a golden staff clasped in the invisible fist of an elderly ghost. For Teiresias the Oracle, Hades was a brighter place than Earth, for he had been blind in the sunlight. Now his grey, cloudy eyes stared piercingly at Odysseus and answered his questions before he asked them. 'Yes, I can tell you what is past and what is to come and what is true. Yes, it is true that there are princes pestering your wife to marry. But she is patient and goes on believing you will return one day. Yes, I can tell you the path you must steer to reach Ithaca. You must sail past the Siren Singers, beside the Clashing Rocks, beneath the lair of the horrible Scylla and past Charybdis, the bottomless whirlpool. Aha! I hear your heart thump even inside the bony cage of your chest. But if you have wisdom enough, you will overcome all these dangers and put in at the Island of the Sun.'

The golden wand of light flickered like a torch flame and Odysseus lost sight of the Oracle's grey face. 'And then? Shall we reach home safely from there? Which way should I steer from the Island of the Sun? Tell me — must any more of my men die? Is Poseidon still angry with me?'

'Angry? He hates you with a hatred deep as the ocean itself. The Scylla will take her fill of men, but their death is appointed for that hour. Do not struggle to save them. Row quickly by. If no one kills or eats the Sun God's cattle which graze on the Island of the Sun, all may be well. All may still be well . . .' The voice faded to a sigh, and the light to a flicker, and the grey face to a wisp of smoke.

Odysseus sprang forward to stop the Oracle leaving, but he slipped on the slime and fell, and a circle of white and doleful faces closed in on him, and invisible fingers felt at his face. The spirits of Hades had forgotten the feel of skin and hair.

Like a swimmer in this shoal of jellyfish, Odysseus flailed his way back to the shore where he had left his fast, black ship. If it had not been for the anxious cluck-clucking of his cockerel, he and his men might never have found its solid hull amongst the softness of the Underworld.

No time for farewells to the dead they knew. No time for questions about life after death. Only a long, sweating pull on the oars, against the current of River Ocean. At last the keel was gripped by a favourable current and emerged into the path of the rising sun. They were swept, without aid of oars, out on to the sunlit sea, and saw Circe's island, a speck on the horizon.

'Magic, merest magic,' thought Odysseus to himself, snuffing up the perfumes of Circe's magic gardens, blown off shore by the sorceress's sighs.

Beauties and Beasts

Circe was overjoyed to see them. She helped them find Elpenor's body and give it burial. His friends planted the rower's oar in his grave mound and called his name three times across the ocean. Elpenor's soul was set to rest for ever.

This done, Odysseus repeated the directions he had been given in Hades, carefully omitting certain details in case his men refused to go on. Circe listened and bit her lip and nodded unhappily. 'If you must go, you must. But since your course lies past the hideous Siren Singers, take beeswax from my hives and stop up your ears before ever you get close to the sound. Once a man has heard the song of the Sirens, his wits fly overboard and nothing can save his soul from shipwreck. Believe me, Odysseus, not even your wisdom could save you.'

Odysseus took the wax. He also promised himself, in his heart of hearts, to hear the Siren song. So when they had put to sea and ploughed a white furrow to the very brink of the horizon, he plugged each man's ears with beeswax and stood beside the mast.

'Polites! Tie me to the mast with rope. And if I ask you to set me free, tie me tighter still.'

'Pardon?' said Polites.

So Odysseus took the wax out of Polites' ears and repeated his instructions. Polites bound him to the mast with a coil of strong hemp, resealed his own ears, and bent over his oar once more.

Across the water came a chirruping like birdsong — an intriguing but not a very beautiful sound. Odysseus strained his ears to hear more. There was no need: the ship passed close by the bald and barnacled rocks where the Sirens sat singing. As it came closer, the singing grew more distinct. It was a song written in an un-nameable key and sung in notes which never climbed the rungs of a musical stave:

> *'Odysseus, see what flowers we have bound*
> *Into a crown for you upon this mound.*
> *A flask of wine and pomegranate sweet*
> *Are waiting here for you to drink and eat.'*

It was true. He could see them. Three women glistening from head to foot with oily balm were beckoning him to come ashore. Their unplaited hair reached as far as the water where it spread out in a fringe of gold around the flowery islet.

'Quick, Polites! Circe was lying. She was jealous, that's all. Just look at those sweet faces. How could they do a man any harm? Put in, Polites! The orders are changed. Put in!'

48

But Polites did not lift his eyes from the deck, and although he cast a quick glance over the rail, his face showed nothing but disgust.

'Polites! I forbid you to row past! Unplug your ears, you fool!' The boat was drawing level now with the island.

'Look, look, my sisters! See his twining curls
— A snare to snare the hearts of we poor girls.
Oh pity us who love you, glorious man!
Put in now! Swim now! Jump now! Come! You can!'

'Polites, cut me free, you fool!' Odysseus writhed until he worked one hand free and could scrabble at the knot binding him. In an instant, Polites and Palmides leapt up from their oars and bound him round, from heels to throat, with a second length of rope. He was all but choking, but he used what breath he had to curse them, to offer them bribes, to threaten them with direst punishments unless they did as he ordered.

The red-prowed boat swept on past the island. Its smell of flowers made Odysseus' head reel. His crew too put their hands to their noses as if the smell was making them dizzy. The sweet song of the Sirens became indistinct and sobbing. 'Ah, let me go, for sweet pity's sake!' groaned Odysseus, straining against the ropes. 'Those poor ladies will be heartbroken if I leave them now!' As the sea fell silent, he slumped exhausted in the cords.

One by one, the rowers unplugged their ears and turned to one another, pulling faces.

'The stench!'

'Those vile creatures!'

'All those bones!'

'All those good men lost.'

'The gods bless Circe for saving us.'

Muttering a thousand apologies, Polites unbound his captain, who was dazed and tearful. 'What do they mean, friend? What stench? What creatures? What bones?'

'Forgive me, my lord Odysseus, but I don't believe you saw those three screeching, scrawny vultures pecking on the bones of a thousand dead and dying sailors. Ah, those poor men — all stretched out like worshippers at a shrine. What a fearful way to die!'

Odysseus nodded, but said nothing. A sprinkling of spray wetted his face, and a noise like distant thunder set the surface of the sea shivering.

Except that it was not thunder at all. It was the Clashing Rocks.

To the port side of the ship, two ridges of rock, razor sharp at the peak, ground together their granite faces like cymbals clashing. The cliff faces gouged and clawed from each other great gouts of spewing fire, boulders and shards which hurtled into

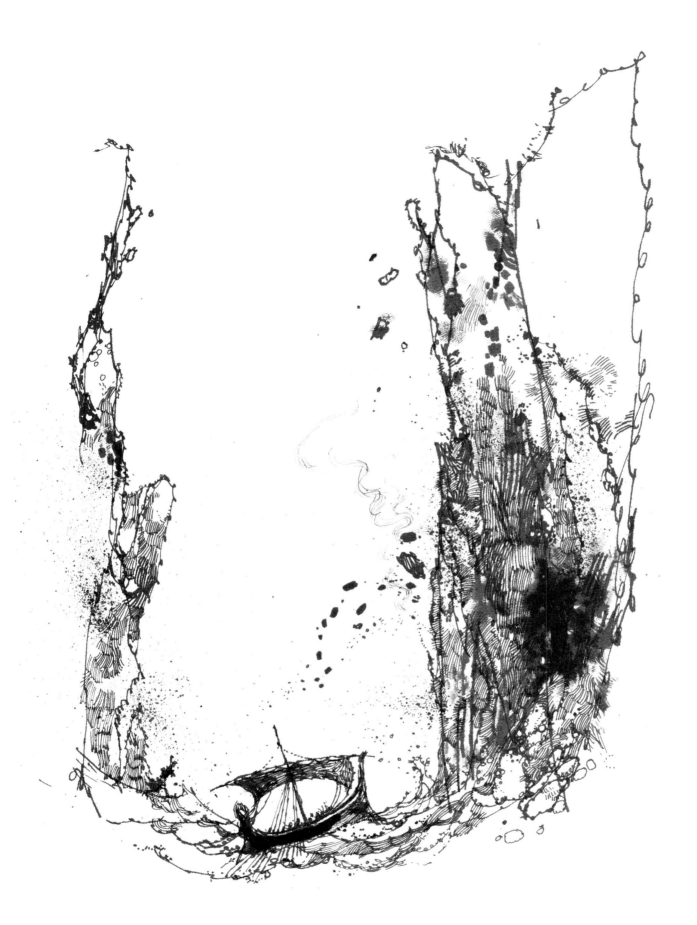

the sea below. The sight and sound was so alarming that the rowers dropped their oars and leapt off their benches to say prayers in the bottom of the boat.

It was all Odysseus could do to remind them, 'You are soldiers and heroes of the Trojan Wars! Pull yourselves together! Besides, if you don't row,' he said calmly, buckling on his sword and setting his brass helmet on his head, 'we may well drift in under those cliffs. Do show some backbone now, or I shall be ashamed to call you men of Ithaca.'

Shame-faced and sheepish, they clambered back to their oars and rowed on. The water bubbled and boiled with the heat of the lava bleeding from the clashing rocks. But though it buckled and bleached the boards of the ship, they were not engulfed by any of the tumbling rockslides as they raced by, muscles straining and eyes fixed on the plume of Odysseus' shining helmet.

He was proud of them — proud till his heart beat fast in his chest. (But he was still careful not to mention what lay beyond the Clashing Rocks.)

The broad ocean was narrowing, narrowing into straits bounded on both sides now by cliffs. To the starboard side a sheer, beetling wall, smooth as alabaster, rose as tall as one of the pillars which hold up Heaven. High up in it, as high as the highest window in King Lamus' palace, a single dark cave overlooked the straits. No path led to it, no Cyclops could come and go with his herd of sheep, the cliff-face was so sheer and smooth.

Of all the men aboard, only Odysseus kept his eyes fixed on that cave. Teiresias' words were branded on his brain: 'Do not struggle, but row quickly by.' All the rest were looking to the other side where, gaping as wide as a harbour and spinning as fast as a chariot wheel, a circle of water whirled in a welter of mist and spray. At the rim, the water heaped itself up, and at the centre it dipped into a spiral, glassy funnel.

Caught up in the maelstrom were the bits and bones of broken boats which had been sucked into the whirlpool, spun to its base, and cracked like eggs against the rocky sea-bed. The noise was like a long, open-mouthed scream, as if all the hurts done to the ocean were being felt in one place.

Twice each day the whirlpool spun to the left; twice each day it spun to the right. Between times, the shining ocean levelled and the whirlpool Charybdis was no more than a clutter of wreckage spinning on the surface. But as the tide ebbed or flowed, the monstrous Charybdis screwed itself, twisted and knotted itself, into a skein of spinning destruction, and sucked in everything that floated on the sea's surface for seven miles around.

As they watched, the whirlpool slowed, slowed and grew shallow. The laughing men shouted their thanks up to Heaven, for surely there would be time to row safely by before Charybdis again breathed in.

Suddenly Odysseus cried: 'Lean on your oars, men! Let me hear your sinews crack! Bend your foreheads to your knees and pull with all your might! And pray, men! Pray as though this were your last day on Earth! Let each man call his name loud enough to be heard in the Underworld!'

Instantly obedient, his men began to call:

'Palmides!'

'Polybus!'

'Eurylochus!'

'Polites!'

'Icmali—ahh! Oh save us, Odysseus!'

No sooner had they called their names, than Içmalius, Eurybates and four more besides were snatched from their benches by the hinged jaws of six serpents.

No, not six serpents but one serpent with six heads — a lizard-backed and scaly beast whose haunches squirmed in its high, cavernous den, while its clawed feet scrabbled down the cliff face and its six heads weaved over the speeding ship. Scylla the monster fed rarely, but well, from the ships which slipped hard-by her cliff-top cave intent on avoiding the whirlpool. Sometimes, when two ships or more were sailing in single file, those following would try to turn back, pushing with all their might against the oars, wrenching aside the tiller. But the pull of Charybdis would still drag them forwards, draw them beneath Scylla's cave, so that she could come a second time and gorge on men or store away future meals in her bone-littered den.

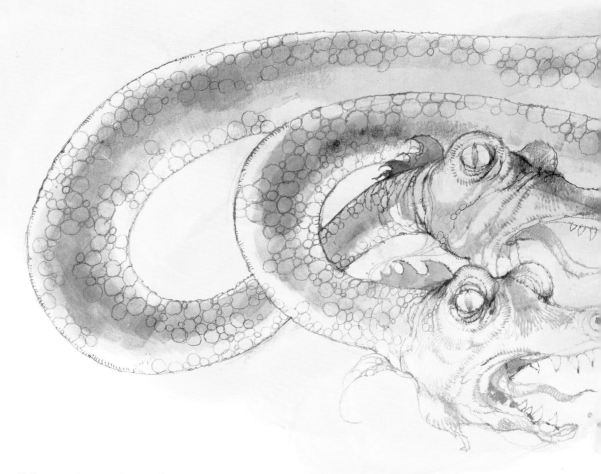

Odysseus knew that only by braving the Scylla's den could those who survived reach home and family: that was why he did not warn the rowers of what was to come. But now he saw hatred in their eyes, because he had steered them close to the monster's cave. Scylla withdrew into her den, and with her went the terrible screams of their six comrades. The rowers had no breath to curse their captain: they were racing against time.

As the six-headed lizard stowed her food, the red-prowed ship leapt forwards — painfully slowly it seemed to claw and wallow its way past the cliff. In panicky fear, the rhythm of the oars was lost and they clattered together and flailed at the air. Scylla re-emerged — each mouth empty, each of her twelve eyes fixed on the little ship. Charybdis, too, began to coil and roar and suck.

With his clenched fist, Odysseus beat out on the prow a rhythm to row by: '*Pull . . .* and *pull . . .* and *pull!*' The sweat ran down; the groans flew up. The Scylla's forepaws scrabbled down the cliff. Her teeth snapped shut — her jaws snatched — and the tillerman felt the breath from two of her twelve nostrils hot on his neck. But they were past her — and past Charybdis, too, though the monstrous whirl of water was gaping wider and wider with every beat of Odysseus' fist on the prow.

58

Mutiny and Murder

Exhausted, they slumped across their oars. Odysseus raised a sail, and a favourable breeze carried them on into the great round O of the central ocean and away from its dangerous, magical margins. The rising moon wounded the sea with a spear of silver, and the old familiar constellations showed themselves one by one like signposts marking the way home.

'Not far now, men. If this wind holds we shall see home within the week. Over yonder, where the sun went down, is the Island of the Sun, but we shan't be putting ashore there.'

And foolishly that was all he said.

Just then, the wind rattled the sail angrily against the mast and the sea shivered into a thousand catspaws. Big warm raindrops hit their weary shoulders as though the gods were spitting on them with contempt. Eurylochus set the boat rocking as he heaved himself to his feet.

'Well, I say we *do* pull in to the Island of the Sun. And I say we light ourselves a fire and find ourselves some shelter and, most of all, I say we get some sleep. I don't know about you, comrades, but my arms have been half out of their sockets and my heart has been half out of my chest with terror. And frankly I don't give a spit for the wishes of a captain who fed six of my friends to the Scylla and never even warned them of the death they had in store!'

Odysseus drew his silver-studded sword and took three paces down the ship towards Eurylochus. But the hands of his other men clasped him round the knees. 'He's right, Captain! We're tired! Zeus knows how tired we are! Why shouldn't we put in at the Island of the Sun? Give us a reason, at least!'

'Death and destruction! Are they reasons enough for you?'

'What? Monsters? Cannibals? Lotus-Eaters? Wolves or bears or Trojans?'

'Cows!' snapped Odysseus peevishly, and suddenly the whole ship burst out laughing.

'Cows?'

'Cows!'

'Moo-hoo-hoo! Ferocious cows!'

Odysseus gave a hiss of exasperation and turned his back on them all and went to the prow. The steersman swung the tiller and the red-prowed ship heeled round towards the Island of the Sun.

They put ashore by the light of lightning, lashed the sail like a tarpaulin across the open boat, and sheltered from the torrential rain. The island itself had no shelter — no farmhouse or ruin, no fisherman's shack or cave or magical villa. It was a miry acre of couch grass and thornbushes. The Cattle of the Sun munched incessantly on the coarse grasses and their long horns clattered together with a hollow, tuneless music. Their red hides streamed with rain, and their velvety nostrils blew bubbles in the pools of rainwater.

Odysseus explained then that Teiresias had forbidden the killing of the Cattle, and the men nodded impatiently. Where was the need to kill the sleek, wet cows? Circe had given them bread, raisins, cheese and pomegranates enough for the voyage. They chewed and baled and baled and chewed and still the rain soaked them through and the wind chilled them. It was a new wind, too.

It was Poseidon's wind. It blew across the Island of the Sun like a razor across a stubbly chin — not towards Ithaca and home but towards Charybdis and Scylla and the Clashing Rocks. The men grew more and more surly. 'You wanted us to keep sailing. We'd be fishfood by now if we'd listened to you!' Odysseus said nothing, but wrung the rainwater out of his beard and stared out to sea.

A week passed: their food was almost gone. The rain still rained and the wind still blew. Another week passed, and Odysseus wrung the neck of his lucky mascot, and they shared one meal of stew. The rain still rained and the wind still blew. Another week passed, and the men's ribs showed through their skin like the frames of sagging tents. The rain still rained and the wind blew harder still. Their hearts faltered and their courage shook, and Odysseus knew that disaster was close at hand.

'What does it matter if we kill the cursed cows for meat?' said Eurylochus finally. 'We're dead if we don't!'

'Don't say that! Don't think it!' shouted Odysseus. 'I'll pray to Pallas Athene: I'll pray to the goddess of war who kept us safe through ten years of battles and hardships around the walls of Troy. Did she give us victory just to let us starve to death now? No! I'll pray to Athene. Just be patient one more day! Look, the rain's stopping even now!' And he left them and leaned into the wind and walked to the other side of the island to pray.

Ever since the killing of the cockerel, he had kept awake, afraid that his men might disobey his orders while he slept. The episode of the bag-of-winds had taught him not to doze off. But he found it difficult to pray, for every time he closed his eyes, the darkness welcomed him like a soft pillow . . .

As soon as he woke he could smell the delicious smoke from roasting beef. He ran headlong, fighting his way through the wind as though it was a heavy curtain hung in his path. Too late. A half-eaten carcass rolled slowly on a spit over a sputtering fire. Not a single crew member had hesitated to cram his mouth with chunks of charred, delicious beef.

He wrenched meat out of their hands and flung it into the surf. But they only
glowered at him and cut themselves some more. Roaring and tearing at his hair, he
fell on his knees and beat his forehead against the red prow. His friend Polites
brought him a rib of beef and knelt down beside him. 'Surely it's better to face death
with a full stomach, my lord. Think kindly of us.'

'I love you all dearly,' said Odysseus, pushing the rib of beef away. 'That's why
I wanted you all to live and see Ithaca again, and your wives and children. And now!
Now, even the cows are mourning our fate!'

A loud lowing — deep and doleful — spiralled past them on the wind. It seemed
to come out of the cloud of smoke which hid the spitted, roasting beef. The men
one by one dropped the meat out of their hands. For the mournful mooing
came not from any of the live cows (which stood in a silent circle around them) but
from the carcass spitted over the fire.

They could not get to the boat quickly enough. They ran it into the water, leaving
behind the burning fire, the delicious smelling meat, some swords and sandals and
shirts and lengths of rope. They bent to their oars like men pursued by monsters, and
they grunted, and ground their teeth with the exertion of rowing. The blades pecked a
broken line of white foam across the sea.

But the trail they left on the greatness of the ocean was no more than the trail of a
snail across the roof of a great city. And beneath them, in the cellars of that city, the
god Poseidon watched their puny progress and smiled. 'I have you now, Odysseus.
You blinded my son the Cyclops, and now I shall plunge your red-prowed ship like a
little fiery stick into the eye of Charybdis!'

Poseidon lifted his head above the waves: he pouched the winds in his cheeks
and crumpled the sea's surface between his two hands. He loosed his white-maned
horses from the icy north, and he caged the rowers round with waterspouts so tall
they seemed to touch Heaven. He planted his feet on the mountain peaks that rise
from the sea floor and he stood, head and chest out of the sea, to shake out his green
hair.

In the darkness they might have escaped. But when sunset came, and the Sun God, passing over his Red Island in the west, looked down and saw the roasted cow, he too swore to be avenged.

'You have slaughtered one of my beasts — my red-backed beasts — my heart's delight! I shall spit you and roast you till you bellow!' And he laid the fiery beams of sunset across the sea and fixed the position of the little ship in the red light, neither sinking nor setting, so that there was no chance of escape under cover of darkness.

Hemmed in by monstrous waves, the little ship was nothing but a straw blowing across the water. A dozen times it stood on its stern and seemed about to plunge, arrow-straight, to the sea-bed. Men were shaken from their benches like olives shaken out of a tree. They plunged into the sea and never resurfaced. The mast fell and carried with it all the ropes, stays, sailcloth and men that clung to it. A wave beat out the base boards of the boat and lifted Eurylochus bodily over the side: he cursed Odysseus as he went under.

And Polites was sucked out through the gaping hole, his hands too wet for Odysseus to keep hold of them.

Soon a noise louder than the howling of the wind and the laughter of Poseidon rose above the chaos: the clashing of the rocks and the roar of Charybdis. In the shadow of the cliff which overhung it, the great wheel of whirling water was just starting its downward spiral.

The hole at its heart grew deeper and deeper by the minute, sucking into it all the floating litter of the slipping sea. Down went casks and kegs. Down went masts and ropes. Over the glassy rim went swimmers, oars and spars. Out over the gaping trough, sheer speed hurled the broken ship. It seemed to pause in mid-air: a single figure could be seen astride its red prow. Then the stern dropped, and it plunged downwards.

The figure on the prow leapt into the plume of spray which hangs continuously over the monstrous Charybdis. He leapt, arms up, into the spray, and he grasped the little thorn tree that grew out of the cliff face. The tree sagged. Its withered roots writhed in the scanty soil that held them. The man's weight seemed certain to wrench it out of the cliff and drop with it into Charybdis.

But Odysseus had not eaten for eight days — not so much as a mouthful of beef. He had never been tall, and his stocky frame was shrunk now to skin and bone and trembling muscle. He knotted his thin legs round the tree and hooked his thin arms over it, and he held as still as the mantis insect that hangs on a blade of grass and waits and prays, and waits and prays.

There under the cover of the spindrift spray, while Poseidon rolled on his back in the ocean trench and laughed, the goddess Athene answered Odysseus' faithful prayers. She reached out an invisible hand and pressed the earth firm around the tree's roots. She cloaked Odysseus from view in a welter of spray, and as the tide turned, and Charybdis' spiral began to unwind, she plucked the red keel of his fast, black ship out of the spinning water.

The whirling water slowed. The whirlpool grew shallow. Odysseus looked down and saw the keel float free of the current. He offered up thanks to the goddess Athene; he unknotted his aching legs and arms and let himself drop astride the red-painted keel. By paddling frantically through the lava-warmed water, he got out of the evil channel — out from between the two dreadful cliffs — and floated back towards the Island of the Sun.

But surely no amount of paddling, no kicking at the water with weary feet, could have saved him from the toils of Charybdis as it began to recoil — not unless some unseen hand had sped him through the water.

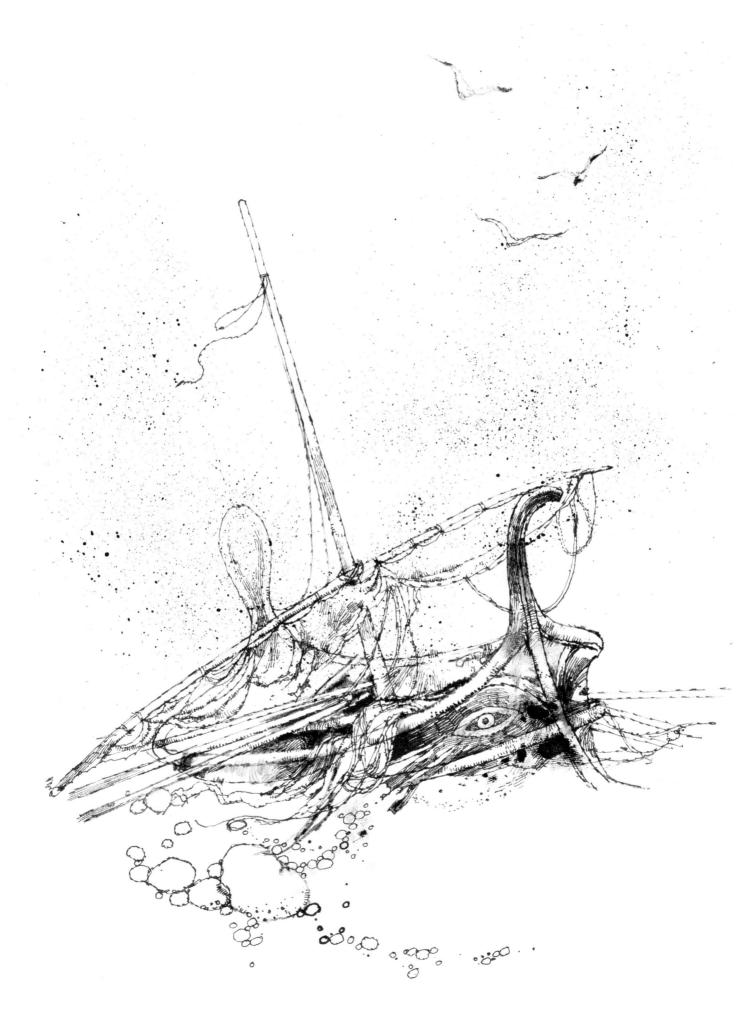

Three Women Watching

A friendly fog smothered the face of the ocean. Neither the vengeful Sun God above nor the vengeful Sea God below sighted Odysseus as he drifted to the shores of Ogygia. There he woke to find himself on a bed of fleeces, with a sea nymph's song in his ears.

Calypso the sea nymph had everything a nymph's heart could desire: hives full of honey, vines bent double with grapes, olive groves and pomegranate trees, figs and carobs and freshwater springs. Her cave was no dark, damp hollow in a rock infested

71

with crabs. It was a sunny cleft high in a flowery hillside. The floor was carpeted with woven rugs seven deep, for each time the colours faded, Calypso had others to throw across them. The walls, too, were hung with tapestries. The weaving of these was her only daily work. Indeed, Calypso wanted for nothing. Nothing, that is, but a husband.

When Odysseus sat up and looked around him, he commented on the prettiness of the cave.

'I have tried always to keep it pleasant while I waited.'

'Without your nursing I would be lying dead on the beach now,' he said. 'I am truly grateful.'

'But how could I let you die after waiting all these years?' she said, and laughed.

'Waiting for what?' He began to feel nervous.

'Waiting for you, of course, Odysseus my love. You are the husband I've waited for all my life. I knew you would come, and now you'll stay with me for ever.'

'But, lady! I'm married! My wife and son are waiting in my three-island kingdom of Ithaca! I must set sail today!'

Calypso narrowed her sea-green eyes. 'But you have no ship, husband.'

'Then you must give me one — lend me one — help me to build one!'

'But I have no ship, husband, and I am the only person living on this island. The trees here are my friends and subjects. They would never allow themselves to be fashioned into a ship that would take you away from me.'

'But my wife . . .!'

'Yes, my dear?'

'My wife Penelope, I mean . . .'

'. . . is old and wrinkled now. I will never grow old: I am immortal.'

Then Odysseus knew how the little bird feels when it lands on lime twigs to rest and finds its feet stuck fast. He threw himself down on the fleecy bed and turned his face to the wall. Calypso smiled patiently. 'Soon you will love me. Wait and see,' she said cheerfully, and returned to her loom which stood by the entrance to the cave.

Far, far across the ocean, in the white halls of Pelicata Palace, another woman was weaving. Penelope, Queen of Ithaca, looked up from her loom and stared out across the wave-striped sea. Every day she sat at her loom in the window of the palace, and every day she watched for Odysseus to sail over the horizon.

But the only ships which came were the ships of smiling, smiling, smiling suitors. Moneyless princes and dispossessed war-lords sailed in on every tide to ask for her hand. The laws of hospitality demanded that she offer them food and drink and a bed to sleep in. But the suitors never went home. They were sleeping four to a bed now. Every day they ate and drank the produce of the island, feasting and making free with Odysseus' own clothes, Odysseus' own weapons, Odysseus' own chairs. Making eyes at Odysseus' own wife!

'He's dead long since, lady. His ship went down in a storm.'

'He was killed by pirates, most likely.'

'Or eaten by cannibals!'

'Don't give him another thought, dear lady. Marry again and give this three-island kingdom of yours a new king.' They talked of love, but none of them loved Penelope: their hearts were set on the golden crown of Ithaca and the island's riches.

'My husband will come back soon,' she told them at first. 'I feel in my heart that he is still alive. He'll be angry to find you here, pestering me. Leave now, that's my advice.'

But as the months went past, she realized that they would not leave just because she advised it. 'Go away!' she told them. 'I don't want to marry any of you. I shall have only one husband in my life and that husband is Odysseus. His son Telemachus will rule the three-island kingdom after him.'

But as the months went past, she realized that the suitors were plotting and scheming to murder Telemachus so that Odysseus, dead or alive, would have no heir.

'Go away!' she told them. 'There's not one of you I'd choose to marry, even if my Odysseus were dead.'

'Then we'll draw lots,' said the suitors. 'The man who wins shall take you for his wife, since you have no special preference.'

Then Penelope asked herself, 'What would Odysseus do if he were in my place? He wouldn't let these bullies have their own way.' And so she set up a loom in the window of Pelicata Palace, and she threaded a wide warp and she wound a heavy shuttle. 'Hear this, you graceless men. I believe now that my husband Odysseus *is* dead. I *will* marry one of you — a man of my own choosing. But not yet. Let me weave a wedding veil and over it weep tears of mourning for my dear, dead Odysseus. When it is finished, I shall choose. Not before.'

That night the suitors feasted more wildly than ever.

'She'll choose me!'

'Never! She's liked me all along.'

'Nah! Haven't you seen the way she looks at me?'

'Why argue? Not long till that veil's woven and we'll know,' they said, and broke open another keg of Odysseus' wine.

That night, when they had all drunk themselves into a stupor, Penelope left her bed and crept to the loom in the moonlit square of the window. 'O moon who shines on me and, somewhere over the sea, shines on my dear Odysseus, too, give me light now enough to do my work.' And she set about unpicking all but a row or two of the weaving she had done during the day. 'Here is one wedding veil which will never be worn to a wedding,' she said to herself. 'Long before it's finished, Odysseus will sail over the horizon and drive these bullies into the Underworld, like sheep into the slaughterhouse.'

But though the friendly moon lit her work and laid a yellow path across the ocean, no red-prowed, fast, black ship sailed along the moon's highway.

High, high above the ocean, another woman sat watching. Like Calypso the sea nymph, and like Queen Penelope, she loved Odysseus, the hero of Troy. She had preserved his life through many battles; she had planted the little white moly flowers which were the antidote to Circe's magic potions. She had pressed home the soil around the roots of the tree which overhangs Charybdis, and she had laid the friendly fog across the sea to hide Odysseus as he drifted helplessly on his red keel. The goddess Pallas Athene loved Odysseus, for all he was small and stocky and mortal.

For seven interminable years she watched Calypso's carpeted cave and saw Odysseus sit sobbing in the sun, pleading with Calypso to let him go. Every day he lit a fire and made sacrifice and prayed to the gods for their help. But every god and goddess on holy Olympus' mountain top had been forbidden to help him. The father of all gods, Zeus the Almighty, had spoken. No meddling goddess was to send a ship or carry Odysseus home on wings of magic.

At last Athene went to her father and said, 'Zeus! Let Odysseus continue on his way. His wife and son need him at home.'

'No!' snapped Zeus. 'He may stay where he is — no torment, surely, to live with a sea nymph in paradise? No, I will not deny Poseidon some revenge for the blinding of his son!'

'Have you heard what Calypso did today?' Athene persisted, her head on one side. 'She offered to make him immortal, as she is.'

'She *what*?'

'She offered him the gift of immortality if only he would love her.'

'The hussy! . . . What did he answer?'

'He refused!' declared Athene proudly.

Zeus gave a sigh of relief and seemed pleasantly surprised. 'Refused immortality? He must really want to leave Calypso very much. What for, I wonder? A wife and a son and a miserable little three-island kingdom?'

Athene waited patiently. 'So you will consider setting him free?'

Zeus scowled, banks of white cloud knitting over his all-seeing eyes. 'When has one little mortal ever caused such excitement among the Immortals? In a few years he'll be nothing but a heap of dust and drifting spirit . . . Very well. I'll send and tell Calypso she must set Odysseus free. But Athene . . .'

'Yes, dearest Father?'

'No magic wings to carry him, no words of advice whispered in his ear while he sleeps, no visits to Earth to walk by his side. The man loves his wife. It would be a great mistake for any foolish goddess to fall in love with him.'

Athene widened her warrior-grey eyes. 'A goddess love a mortal? How could such a thing ever happen, Father dear?'

Behind a rainbow column of the House of the Gods, Poseidon crouched, greenly out of place in the dry realm of Heaven. He dripped wet blue rain on the landscape beneath as he rubbed his hands together and bared his teeth in a smile. 'Now I shall have you where I want you, little mortal. Soon you will wonder why you ever prayed to leave Calypso's isle!'

Poseidon's Revenge

Zeus' messenger was an unwelcome visitor to Calypso. She wept, she stormed, she pleaded, but at last she had to submit and let Odysseus go. She allowed her trees to be felled and bound together into a raft, and she even wove a sail to hang from its mast. But all the time she coaxed and wheedled: 'Don't you love me just a little? What don't you like about me? I'll change! You could love me if only you would make the effort. I'd make you immortal. Don't you want to be immortal? Do you want to die one day and go down into the Underworld for ever? Do you want to go out there and face Poseidon? He'll remember you! He'll never forgive you!'

'Madam, I am very grateful to you for saving my life,' said Odysseus, straining to push his raft into the water. 'I shall certainly remember you for the rest of my life.'

'Will you? Oh, will you really?' She stood on tiptoe in the shallows, her hand peaked over her eyes, and waved him goodbye until his sail was no more than a white speck on the horizon. 'He would have loved me, if only he had stayed another few weeks,' she said to herself. Then her attention was caught by a fork of lightning which stabbed the northern ocean.

A fork? It was a trident — the golden trident Poseidon brandishes to quell the rebellious sea beasts and tame the shark. Out of the east came a herd of sea horses, arching their foaming white manes and trampling the sea into a dented and buckled grey. Wrecked galleys and fishing boats, which had lain empty and broken on the sea floor, were scooped up now and hurled across the water.

By the light of lightning bolts which rained down around him, Odysseus saw the frightened, colourless eyes of fishes, and the suckered arms of reaching squid. The waves that folded over him were shot through with eels and peppered with sharp barnacles and razorshells. The troughs that swallowed him were deeper and darker than Charybdis, and the currents beneath dragged him three times round the ocean like dead Hector was dragged three times round the walls of Troy.

Then the barbs of Poseidon's trident set the ropes alight which bound together Odysseus' raft. The logs floated apart, smouldering, and Odysseus was thrown into the gnashing sea and swallowed, body and soul.

He shed his sandals and heavy skirt and warrior's sword. They dropped away beneath him into the bottomless dark. He held his breath until he felt something wind around his chest that he took for an octopus. And only then did he despair and breathe in deeply and fill his lungs with water so as to be quicker drowned.

81

The water tasted like air — like sweet, fresh air! And no sea monster had hold of him. Something pale and smooth brushed against him, but it was only a girl with a fish's tail and long strands of seaweed hair. Her scarf was round his chest, and she towed him playfully by its ends, her cheek pressed close to his. When they broke surface, the lightning illuminated a fearful reef on which the storm waves were dashing themselves into glittering clouds of spray. Odysseus too seemed bound to be broken against the razory rocks, and yet the girl continued to wriggle and giggle and tow him about in the loop of her magical scarf.

The night was pale with weariness, but the sun was not yet up. The land beyond the reef came gradually into view, detail by detail — a gap in the reef, a steep hill, a river estuary.

Suddenly, the sea nymph's game was over and like a child tiring of a toy, she swam off, pulling her scarf behind her. Without its magic he was once again spluttering and floundering, half drowned by each ferocious wave. Only weary and desperate swimming brought him at last into the river. On hands and knees he crawled up the icy watercourse before pulling himself ashore and into a low tree: he was afraid of wild animals eating him from head to foot before he could even wake up.

In the event, it was not a wild animal which woke him but a troop of girls come to wash and do their washing at the river. One took off her gown and, not seeing him, accidentally draped it over his face in hanging it from the tree. He woke in a panic, dreaming he was back in Calypso's cave, being smothered by woven carpets. In fighting off the gown, he fell out of the tree and into the river with a loud splash.

When he spluttered to his feet, he was surrounded by young women submerged up to their shoulders, like him, and all staring with round, startled eyes. 'How dare you, sir!' said the tallest. 'Were you spying on us?'

Odysseus shook the water out of his ears. 'Certainly not, madam! I don't mind if I never see another young woman in all my life. They only make for trouble. Falling in love with me and so forth.'

'I can't imagine why,' said one, with profound scorn.

'Caliope, hush! The laws of hospitality demand that we should be polite to this fellow, no matter what. Do you have a name, old man?'

'Old man?' Odysseus' jaw dropped. He scrambled out of the river and picked up a mirror which belonged to one of the girls. He did not recognize the face he saw in it. It was wrinkled and chapped by the sun and sea water. The beard and hair were grey, the eyebrows caked with salt, and the eyes red-rimmed and bloodshot. It dawned on him, too, that he was dressed only in his shirt, and that his bony, fish-nibbled knees were knocking together. He dropped the mirror. 'Ladies! How can you ever forgive me for my behaviour? How can I ever make you believe that I am Odysseus, King of Ithaca, returning from the Trojan Wars?'

Then all the girls but one burst out laughing. The tallest said, 'If you were to turn your back while we got out of the river, we might just believe you were a gentleman!'

An hour later, Odysseus was riding in the back of a cart, in among the wet washing of the Princess Nausicaa and her waiting-women as they drove back up to the palace of King Alcinous on the island of Scheria. And there Odysseus presented himself — a humble, nervous, shame-faced, worn and weary man in a torn and dirty shirt.

King Alcinous was a man with treasure houses and armouries, a thousand acres of farmland and a fleet of red-prowed ships rocking in a stone-built harbour. His household numbered a hundred servants, and his temples made ceaseless sacrifice to the gods. His merchant ships, criss-crossing the oceans, carried the King's fame as far as Africa and the Pillars of Hercules.

But when he saw small Odysseus, bent and ragged and covered in saltwater sores, he got up from his place at table and took him by the shoulders. 'You told my daughter Nausicaa that you are Odysseus, King of Ithaca, and I see in your eyes that you spoke the truth. Sit down now and eat and drink. I shall have fresh clothes brought for you and a ship made ready and filled with a few humble gifts. When you are rested, if you can bear to tell us your adventures, we would be most privileged. Your name is famous from shore to shore of the world's central sea.'

Then the ragged King of Ithaca was moved to tears. He embraced Alcinous. 'I will tell you everything and miss nothing out,' he said, drying his eyes. 'But first tell me one thing. If you have heard of my name, do you know of my little three-island kingdom? It's called Ithaca, you know, and I would dearly like to see it again.'

'But of course I know it, dear friend! Wooded Zanthe is just over the horizon, and beyond that is Cephalonia and beyond that Ithaca with its towering Mount Neriton. A day's rowing by my best men will bring you safe home to your lovely queen.'

The banquet that King Alcinous mounted that night was sung of by poets and bards in songs and ballads until, in due course, it echoed off the Clashing Rocks, rang in the sea caves of Calypso, rose up to the ears of the gods and was swept by River Ocean into the shadow of the Underworld. Between courses of food and between the dances of dancing girls and the playing of musicians, Odysseus told his adventures.

Telling them was rather like reliving them, but he missed nothing out — how and why his friends and companions had died one by one, and where his voyages had taken him. The ladies hid their faces when he described the monsters. Grown men wept when he described the loss of twelve ships with all their crews. A whole night it took to tell the whole story — and yet the whole of it was not yet told, for still Odysseus was separated from his wife and son and three-island kingdom.

So at dawn he boarded the ship which Alcinous gave him, and lay down on the foredeck, on a pile of blankets and embroidered clothing which Princess Nausicaa herself laid ready. The hold was filled with copper cauldrons, chests of silver, and gifts of linen and perfumes for Penelope. Every young oarsman in the Scherian navy wanted a place in the ship, and the crew was chosen by drawing lots. Never a word was spoken about Poseidon or his unsatisfied revenge.

For ten years Odysseus had slept no more than a bird on the wing. Now, after a night's storytelling, he slept so deep and dreamlessly that he did not wake even when the Scherian ship beached in a gravel cove — even when the rowers carried him, foredeck treasure and all, out of the ship — even as they heaved their red-prowed ship back into the surf and rowed away singing songs about the ten-year voyage they called the Odyssey.

They were almost home. They sang in time with the beat of their oars, and the song filtered down through the water and set the green strands of Poseidon's hair swaying. The Sea God, who had been sleeping, stretched out in the sea's deepest trench, roused up, and listened with only half an ear.

'Odysseus is home again —
Safely home with wife and kin!
We his oarsmen share his fame:
Name us when you tell of him!
His voyaging is over now.
Pushing through Poseidon's sea
See! our crimson-painted prow
Ends the hero's Odyssey!'

Poseidon's roar set the sea boiling. His head broke through the waves not a stone's throw from the ship, and his hands encircled it like green Charybdis.

Such was the phosphorescence that shone around him that the fast-ship of Scheria and all its crew were engulfed in light. Such was the look in his eyes, scowling at them across the scarlet prow, that their hearts turned to stone with the weight of terror.

Not only their hearts turned to stone: the chests which cradled them, the legs braced against the base boards, the base boards themselves; the arms which were hauling on the oars, the oars themselves. The whole ship, from prow to stern, was turned to stone — even the water which had buoyed it up — so that it stood on a stem of rock: a black islet flaking red paint into the sea.

In future years, sailors passing it would offer up devout prayers and sacrifice to Poseidon, but murmur under their breath, 'Hail to the brave men of Scheria and to King Alcinous whose kindness is commemorated for ever by this sad, black rock!'

A Husband for Penelope

On the night that Odysseus told his adventures to King Alcinous, his Queen Penelope was also awake, working at her loom in the moonlit window of Pelicata Palace. She, too, was weary after years of sleeplessness — all those lonely nights spent unpicking the threads she had woven during the day. She was determined that the veil must never be finished.

The suitors had begun to wonder, years before, why the veil was so long in the weaving. Most thought that some magic force must be unravelling the work to spite them. But two were not superstitious at all. They kept a watch on Penelope while she worked, waiting to catch her out unthreading when she should be threading. They were baffled.

Now a new idea came to them.

If Penelope had not been so weary, nodding over her moonlit work, she might have heard their clumsy footsteps on the stairs and their creeping in at the door. Suddenly they burst into the room and picked up the loom and pitched it through the wide window.

'You are found out, madam! Your deceit is uncovered! Who'd have thought it? You're as cunning as that dead husband of yours, the wily old Odysseus. Well, the game's over. The veil's finished. Today's the day you'll choose a husband, lady, so you'd best take a close look at each of us. Choose me and I won't tell the others what a sly deceiver the Queen Penelope is.'

'But I will!' said his companion peevishly. 'I mean to have her before you!' And they went away, arguing and bickering.

'Do your worst!' called Penelope after them. 'I am Penelope, daughter of Icarius and wife of Odysseus! How could I agree to marry any of you? Not one of you would make so much as a fit sheath for Odysseus' sword!'

The word quickly spread of Penelope's ruse, and the suitors emptied the larders in preparation for one last magnificent feast at which the marriage would be settled once and for all.

When Odysseus woke, lying on the heap of embroidered garments, he could not, for a moment, remember where he was. The shape of the mountain which towered over him was somehow familiar.

'Mount Neriton!' He was home on Ithaca — alone on a gravel foreshore, with no one to thank for his safe return and no one to greet him either.

Fear shook his heart when he thought how long it had been since he learned of the suitors beseiging Penelope. Could she possibly have gone on believing in him, waiting for him, fending off the advances of the ruthless princes? Surely by now she had been forced to marry. He shuddered at the thought, then quieted his beating heart and thought of a plan.

Hiding the treasure given him by King Alcinous, he put on again the tattered, filthy shirt in which he had been found by Nausicaa. He dirtied his face, wrapped his head in a piece of old sacking, and climbed familiar pathways to the home of an old friend.

The ancient pig-man still lived in his bare, uncomfortable cave, tending the palace pigs as he had when Odysseus left for war. There were few pigs left, though the herd had once been huge: the suitors gorged themselves daily on pork and bacon. As Odysseus approached the cave, he heard the drone of voices — a young man and an old one — and his own name was mentioned more than once . . .

'It would be different if Odysseus were here,' said the old pig-man.

'Would it? Would it? All my life I've heard so, but Odysseus is nothing but a name to me. I don't even remember his face.'

'Why, lad, you only have to look in a mirror to see what Odysseus looked like. You're the picture of your father.'

Odysseus ripped aside the ragged curtain across the doorway of the cave. 'Telemachus? Is this really Telemachus?'

The young man leapt to his feet, half drawing his sword, thinking to be ambushed (as he had been ambushed before by the suitors). The pig-man leapt between them. 'Aha! I know that voice!' he said, peering into the newcomer's face with a big winking grin. 'No need for the sword, Prince Telemachus. This is an old friend of mine. He's been away travelling the world for a great many years.'

Odysseus winked, too, at the old man and said, 'And now I'm shipwrecked on your beautiful island and have no means of getting home. Would you help me to a new boat, Prince Telemachus, even though I'm a stranger to you?'

Telemachus gave a snort of disgust. 'You certainly have been gone a long time if you don't know the state of things here on Ithaca. My word counts for nothing. After today I shall be lucky if I keep my skin. Now, if my father were here, he would give you a boat and everything you need to reach home. He's a traveller himself, and must need the help of strangers.'

'Who? Odysseus who fought at Troy?' said Odysseus. 'Did he never come home, then? Perhaps he's dead.'

'Then the Queen and I and Ithaca are lost. I *won't* believe it . . . You know, your face is familiar. Have we ever met before?'

Odysseus put one arm across his eyes to hide his tears of joy. 'Not since you were a newborn baby, not for twenty years: not since the Trojan Wars began and every true man left his home and family and went to fight in the service of Agamemnon. Oh! you can't imagine how hard it was to leave my wife and baby — or the trouble I've had in returning to them. Come here, son, and let me look at you. I am your father, Odysseus. I have come home at last!'

Telemachus left the pig-man's cave before Odysseus, and returned to the palace just as if nothing had happened. He said nothing to his mother, nothing to the suitors who jeered at him as he came in and jostled him with their elbows. The feast was ready at which Penelope must choose her new husband.

The suitors no longer tried to please or flatter her. She was now simply a prize one of them would win, and the wedding simply an excuse to eat and drink all they could lay hands on.

The suitors hooted and laughed and drank so much that no one noticed a threadbare beggar creep into the yard and sit down by the door. No one, that is, but a big old dog lying out in the last heat of the sun. Its ribs were bruised by kicks from the suitors, and it swayed unsteadily on painful hips. But at last it reached the beggar and snuffed up his many smells. Then it laid its head in the beggar's lap and its tail thumped the ground three times.

'So you remember me, do you, Argos, my faithful old friend,' said the beggar. 'You remember how we used to go out hunting together when you were just a silly young puppy. We've had a hard life since then, you and I. What a lot of things we could tell each other, eh, old boy?' And he fondled the dog's ears until the faithful creature's heart burst with joy and he died in the beggar's lap.

After a few minutes, the beggar lifted the shaggy head aside and entered the hall, bowing and creeping most humbly. He knelt beside each chair in turn. 'Spare me a little meat from your plate, sir,' he said.

'Get outside with the other animals.'

'Spare me a sip of wine from your cup, sir.'

'What? And drink from it after? I'd catch something. Get away.'

'Spare me that crust of bread in your fist, sir.'

'I'll spare it, yes,' said the suitor, and threw it in the beggar's face, then hurled apples and lemons against his back as he crept away.

'Spare me a bite to eat, lady, and I'll remember you in my prayers.'

'Here, sir. You may have my dinner and my wine,' said Penelope. 'It would choke me to eat in the company of these uncharitable dogs. Here, sit in my chair and rest yourself. I pray that somewhere someone has meat and drink to spare for my dear husband.' She rose to leave the room, but when the suitors caught sight of her they set up a roaring:

'Where are you going? You can't go yet! You've not chosen! Choose!'

'Choose!'

'Or shall we choose for you?'

'Tomorrow you'll be sitting down to dinner alone with your new husband!'

White faced, Penelope silenced them with a glance of her piercing eyes. She drew herself up to her full height and seemed about to refuse marriage one last time.

'Yes! Choose, Mother!' cried Telemachus, jumping to his feet. 'It's time you did. Obviously my father is dead. How could anyone spend *ten* years coming home? Choose, Mother. I was once heir to Ithaca, but I don't care any more: let one of these noble gentlemen have the crown, and his sons after him.'

'Well said, boy!' crowed the suitors. 'At last he's grown up!'

Queen Penelope was dumbstruck. 'My own son tells me this? Then I give up.' She added bitterly, 'Since you think I should give myself away into the hands of these men, Telemachus, perhaps you should choose which one I marry.'

'Let them compete,' said Telemachus quickly. 'Since your first husband was a man skilled with weapons, why don't you marry the one who can match Odysseus in skill? Look! There's father's bow still hanging over the fire. Marry the man who can string it and shoot an arrow and hit a target of your choosing.'

Forlorn, desolate, and betrayed, Penelope searched about for the most difficult target she could name. Each suitor carried a cleaver or axe swinging from his belt by a leather loop. 'Set your axes head-down on the table, and let the man who strings the bow and fires an arrow through the belt loops make me his wife tomorrow morning.' And she swept out of the hall and went to her bedroom.

With a drunken cheer, the suitors swept the dishes off the table, and kicked aside the beggar who was sitting in the Queen's seat. They slammed down their axes and they clawed the bow off the wall — the great hunting bow of young King Odysseus. One by one, they strained to bend it so as to slide the string's loop into the cleft at the tip of the bow.

They grunted and struggled. They swore and they failed. Each man that gave up threw the bow away from him in disgust. 'It's impossible. It's gone stiff with age. It would take the strength of three men to bend it!'

'Let me try,' said the beggar, who had sat silently all this while.

'You, you piece of dirt?' Again they showered him with fruit and kicks.

'Let him try,' said Telemachus, scornfully tossing the bow at the beggar's head.

The tattered man stood up — not a tall man, but broad-shouldered and stocky. He leaned the bow across one thigh and braced it behind the other ankle, then slipped the string loop into its cleft. The bow was strung.

Blustering with rage, one of the suitors snatched it away. 'Well, let's get on with the contest, then! Me first!'

The axes tumbled, the arrows shied left and right until the walls of the hall bristled. They were failing, failing, failing. They were furious at failing. They were wild with disappointment. They hated one another in case one at long last succeeded. They threatened Telemachus with clenched fists because he dared to laugh at their miserable efforts.

The beggar did not laugh. He waited until the bow was thrown aside by the last unsuccessful suitor. All the axes were standing. He took aim through the dozen leather loops and fired.

The arrow plunged through the loops as straight as light through the iris of the eye — and pierced a suitor in the heart. After that the beggar leapt on to the middle of the table, amid the forest of axes. His head and chest were bare, his grey hair curled to his shoulders. 'I am Odysseus, home from the wars of Troy, and you are the ants I found in my larder, the rats I found in my cellar! Penelope will marry none of you. She has a husband already, as you will soon regret!' He loosed a dozen arrows. Every one found its target. His son leapt up beside him, with two swords, and back-to-back they fought.

Against boy and against beggar the suitors had been brave enough. Against Odysseus and the heir to his crown they fell into a panic and squealed and stampeded like men transformed by magic into pigs. But there was no escape. An hour later the room fell silent. Every suitor lay dead.

Telemachus sat down in the middle of the table to catch his breath. But Odysseus touched him shyly on the shoulder and pointed in the direction of Penelope's room. 'You go and tell her I'm home. I don't know how.' Telemachus went and told her.

When Penelope came to the head of the stairs, her face was unsmiling. She bowed her head to Odysseus and waved him towards a chair. 'You must be weary, sir, after all your travelling. I am most grateful to you for ridding the palace of those wasters,' and she waved her hand about her at the pile of dead bodies. 'I shall prepare a bed for you.'

It was Odysseus' turn to be dumbstruck. So cold a welcome after twenty years? Well, perhaps he was not the handsome husband she had sent away to war. Perhaps he was a disappointment to her. 'Could I not sleep in my own bed?' he asked timidly.

'Very well, I shall have it carried to the West Room. You will be comfortable there.'

Odysseus clapped his hands. 'Now I understand! You are testing me, lady! My bed is carved out of the strongest branch of the tree which stands in the centre of this house and which holds up its roof. How could it be moved "to the West Room"?'

Then Penelope leapt across the dead suitors on the floor and kissed her husband and held him close. 'After so long, I had to test you — I didn't dare to believe my own eyes. I expected to see an old man worn out by struggles and hardships. But you're just as handsome as the day you left Ithaca!'

In the fields of the three-island kingdom the farmers danced. On the slopes of Mount Neriton the goatherds played their squawking pipes. Throughout Ithaca and Cephalonia and wooded Zanthe, beacon fires were lit and drums were beaten from morning till night to say that King Odysseus was home at last.

In the peaceful days that followed, poets wrote down the griefs and triumphs of the Odyssey. But Odysseus would not hear the poems recited nor the songs sung, until he had made sacrifice to Poseidon and struck a peace between man and god, between sea and land, between Heaven and Earth.

Oxford University Press, Walton Street, Oxford OX2 6DP
Oxford New York
Athens Auckland Bangkok Bombay
Calcutta Cape Town Dar es Salaam Delhi
Florence Hong Kong Istanbul Karachi
Kuala Lumpur Madras Madrid Melbourne
Mexico City Nairobi Paris Singapore
Taipei Tokyo Toronto

and associated companies in
Berlin Ibadan

Oxford is a trade mark of Oxford University Press

Text copyright © Geraldine McCaughrean 1993
Illustrations copyright © Victor G. Ambrus 1993
First published 1993
First published in paperback 1996

A CIP catalogue record for this book is available
from the British Library

ISBN 0 19 274130 6 (hardback)
ISBN 0 19 274153 5 (paperback)

Printed in Hong Kong